SOS

SOS

THE STORY OF
RADIO-COMMUNICATION

G. E. C. Wedlake

**CRANE, RUSSAK & COMPANY, INC.
NEW YORK**

© G. E. C. Wedlake 1973

All rights reserved. No part of this publication may be reproduced, stored in a retrieval system, or transmitted, in any form or by any means, electronic, mechanical, photocopying, recording or otherwise, without the prior permission of the publishers

Published in the United States by:
Crane, Russak & Company, Inc.
52 Vanderbilt Avenue
New York, N.Y. 10017

Library of Congress Catalog Card Number 73-91529
ISBN 0-8448-0270-0

Printed in Great Britain

*To
Florence May
and
Jennifer May*

Contents

List of Illustrations		9
Preface		11
1	Experimental	13
2	Mushroom Growth	30
3	For Those in Peril	49
4	Rival Systems	63
5	Trans-Ocean	74
6	Preliminaries to War	88
7	World War I at Home and at Sea	100
8	World War I on Land and in the Air	116
9	Progress between the Wars	130
10	Words and Music: Part One	147
11	Words and Music: Part Two	161
12	Television	182
13	World War II	188
14	The Eagle Has Landed	199
Appendices		
	A: Particulars of Early Instruments	207
	B: Abbreviations	222
	C: Chronological Table	223
Selected Bibliography		232
Acknowledgements		233
Index		234

List of Illustrations

PLATES

Early Marconi coherer receiver	33
Early Marconi wireless telegraph transmitter	33
Poldha transmitting apparatus	34
Earliest form of mobile radio (1901)	34
A still from the first cinematographic film about wireless	51
Wireless in World War I	51
Small portable DF station (about 1913)	52
Bicycle power unit of World War I	52
Trench WIT station on Gallipoli in World War I	85
Marconi ½RW pack set at Falahiyeh, Mesopotamia, in World War I	85
American soldier operating a trench set on the Western Front	86
Ship's radio-room	86
Replica of wireless cabin of 1920 at the Marconi Marine Jubilee Exhibition, Baltic Exchange, 1950	103
Marconi type 12A direction finder (about 1920)	103
World's first wireless telephony news service	104
Marconi mobile transmitter/receiver, complete with power generator	104

FIGURES

Samuel Morse's experiment for passing electric signals between two points without any metal connection	16
Induction system	18
First land stations, 1901	38
First US naval stations	67
Marconi directional aerial system	78
Position of blockading fleets and of DF stations monitoring movements of German fleet	111
Beam homing system	136
Long-distance transmission via satellite	204
Induction coil	207
Hertz's transmitter and receiver	209
Marconi's first transmitter	210
Marconi's first receiver	210
Improved Marconi transmitter	211
Improved Marconi receiver	212
Lodge transmitter	212
Goldschmidt radio frequency alternator	214
Poulsen arc transmitter	215
Multiple tuner and magnetic detector	216
Electrolytic detector	218
Physiological detector	219
Radio-sonde system for meteorological observations in the higher atmosphere	221

Preface

Rather less than one hundred years ago Heinrich Adolph Hertz showed, in what must surely be regarded as one of the most important scientific experiments ever carried out, that the invisible, intangible waves, which a few years earlier James Clerk Maxwell had described on paper, did in fact exist.

Tragically, Hertz died young, before his tremendous discovery had begun to bear fruit, and it was left to others, of whom Guglielmo Marconi was one of the first, to show how the 'Hertzian' waves could vastly extend man's control over his environment, first enabling him to send messages over long distances, then to span oceans with voice and music, and finally to venture into outer space.

What follows is an account of the development of radio communication. It is a story full of drama and surprises. To follow it no special knowledge is needed and only a few simple technical explanations are included in the text. However, for the knowledgeable there is an appendix in which will be found described some of the once important but now long-forgotten gear; there is also a chronological table which reveals how old some 'new' ideas can be.

G.E.C.W.

I

Experimental

From earliest times man has had a need to communicate with his fellows beyond the range of human voice or sight, to give warning of danger, to call for help, to summon a council; but until comparatively recently his means of doing so were limited. Signals with a simple, prearranged meaning could be sent by smoke; beacons—the method said to have been used to announce the fall of Troy—were employed; and there was the quite sophisticated method devised by Alexander the Great, who anticipated the nineteenth-century heliograph by using a burnished shield to reflect the rays of the sun. But more elaborate messages could be conveyed only by horseman, runner or carrier pigeon.

Coming much nearer to modern times, flag signalling was introduced at the end of the eighteenth century for signalling between ships, and during the Peninsular War an elaborate semaphore system was developed for communication between the British army and the fleet. However, until the third decade of the nineteenth century the horse remained the basis of rapid communication between points on land.

The position was entirely changed by Michael Faraday's discoveries in the field of electrical science. These opened up vast areas for development and in particular made possible the devising of an electrical system of signalling. A number of scientists set about discovering such a system, the first to succeed being an American, Samuel F. B. Morse; Morse had been a professional artist until he was forty and then had turned scientist, and it is he of course who has given his name to the signalling code now in general use. In 1835 he produced the first practical electric telegraph, in which signals were transmitted from point to point along a copper wire. The first public service using his system was opened between Washington DC and Baltimore in 1841 and it was not long before a network of wires linked every important city in the major countries.

The coming of the telegraph had an immense impact on the commercial and social life of the whole civilised world. For the first time it was possible to initiate and conclude a deal over a long distance in a matter of hours; for the first time too people could summon a distant relative to a sickbed, could announce a happy event, could even back a horse on a distant racecourse; and all at a cost within the means of all but the very poor.

The introduction of submarine cables was delayed for a few years pending the development of suitable gutta-percha insulation. But at last, in 1851, one was laid between Dover and Calais, for the first time connecting two places separated by the sea. This was followed by the first successful Atlantic cable in 1866, and a cable from England to Australia in 1871. Now the nations were no longer isolated from each other, and any event taking place within the limits of the international telegraph and cable network had immediate repercussions in many distant lands.

In the world of commerce the impact of this new power of instant communication was revolutionary. Previously it had taken weeks or even months to receive a reply from a foreign agent or supplier, and information from distant markets could well be out of date before it was received. Now communication took no more than hours. What could be called world market prices came into existence and these were known immediately in all commercial centres, as was news of any sudden crisis which might affect them.

The change was particularly marked in the field of shipping. Before the coming of the cables a ship, fully loaded, would sail from her home port for a foreign destination, but it would often be difficult or even impossible for her owners to arrange a cargo for her return voyage. The result was that the captain had to act as the owner's agent for this purpose. Inevitably this led to delays, with ships often lying idle for weeks instead of, as today, being sent off as soon as they finished discharging in one port to some other where there was a cargo for them. The cable made it possible for a shipowner to be informed as soon as a ship was sighted, as well as when she arrived in port, and he was then able to send the necessary orders to the captain.

There was, though, one qualification to this—and it was a big one—implicit in the phrase 'when she arrived in port'. Once she had disappeared over the horizon the owner had no precise knowledge of the ship's whereabouts or of when she could be expected to arrive at her destination, nor indeed, if she turned out to be late in arriving, whether she was still afloat. This drawback related to the one serious limitation in Morse's telegraph: it only worked when the transmitting and the receiving instruments were linked by a copper wire—in other words, only between two fixed points. One end of the wire could not be fixed to a ship at sea, to a body of men on the march, or to any other moving point; still less then could the wire link two ships.

This limitation was very soon realised, and throughout the second half of the nineteenth century scientists studied the problem of how to dispense with the copper wire link. It is with the means found to cope with this problem—first on a scale that by present-day standards seems tiny, and eventually on one immeasurably greater than even the most sanguine of the early pioneers can have dreamed—that we shall be dealing in this book; we shall also be studying the tremendous impact which the various developments arising from those technical solutions have had on almost every facet of human life.

It was Samuel Morse himself who in 1842, only seven years after the announcement of his invention of the line telegraph, succeeded for the first time in passing electrical signals between two points without the aid of any metal connection. Carrying out his experiment on a canal in Washington (Fig 1), on one bank he placed, a few yards apart, two metal plates in the water, connecting them by an insulated wire through a battery and a telegraph key. On the other side of the canal he placed two similar plates in the water, this time connecting them through a galvanometer. When the key was pressed the galvanometer needle was deflected.

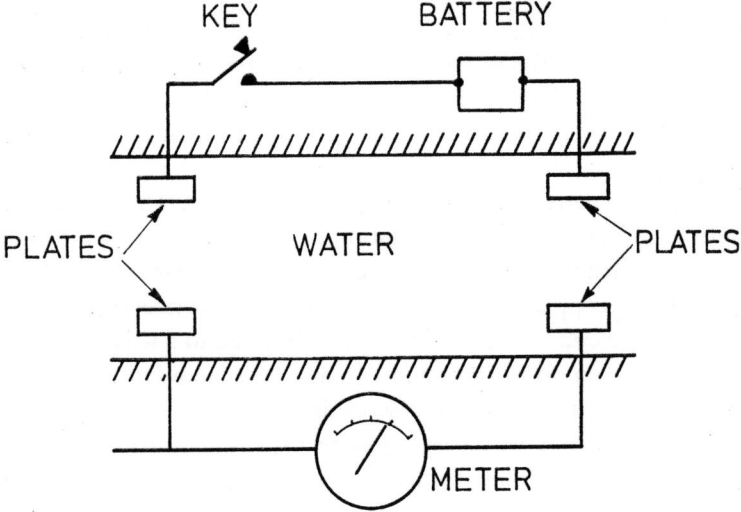

Fig 1 Samuel Morse's experiment for passing electric signals between two points without any metal connection

This became known as the conductive system of wireless telegraphy and numerous eminent scientists experimented with this system until nearly the end of the century. J. B. Lindsay of Dundee invented the method independently, carrying out experiments on the River Tay from 1845 onwards, and later Professor John Trowbridge of Harvard found that signals could be sent by this method for a considerable distance through the earth.

Experimental

In 1822 Graham Bell, one of the pioneers of the telephone, carried out an interesting experiment on the River Potomac in the United States. Using the conductive method, he succeeded in communicating between two vessels lying a mile and a half apart, by means that could fairly be called 'wireless' telegraphy. A little later Sir William E. Preece used the same method to signal across the Solent.

It was found that the method could be used for telephony as well as telegraphy and in 1887 Professor A. W. Heaviside succeeded in communicating in speech from the surface to a depth of 350ft in a coal mine. Experiments with the system continued until 1895 or so, but the distances that were achieved were never sufficient to make it of practical use. However, during World War I some interest in it was revived. Numerous difficulties were then being experienced in operating wireless in forward positions on the Western Front, not the least being that masts and aerials were easy targets. Earth conductive telegraphy, as it was then called, needed none, and for this reason it seemed attractive. But suitable instruments were not available and the matter was not pursued.

However, to retrace our steps a little, it was in 1891 that Professor E. Trowbridge introduced the inductive method (see Fig 2). The transmitter consisted of a large loop of wire (C^1) across the end of which a battery (B) and a key (K) were connected. The receiver consisted of a similar loop (C^2) across which was a telephone (T). When the key was pressed a click was produced in the telephone. The two loops, in fact, formed the primary and secondary of a very loosely coupled transformer.

Sir Oliver Lodge also experimented with this system, and he described his results in an article published in the journal *Nature* of 20 February 1890. He found that by introducing condensers into the receiving circuit he could 'tune' it, so establishing resonance with the transmitting circuit. This not only enabled the power required to transmit over a given distance to be substantially reduced, it also made selective tuning possible. It

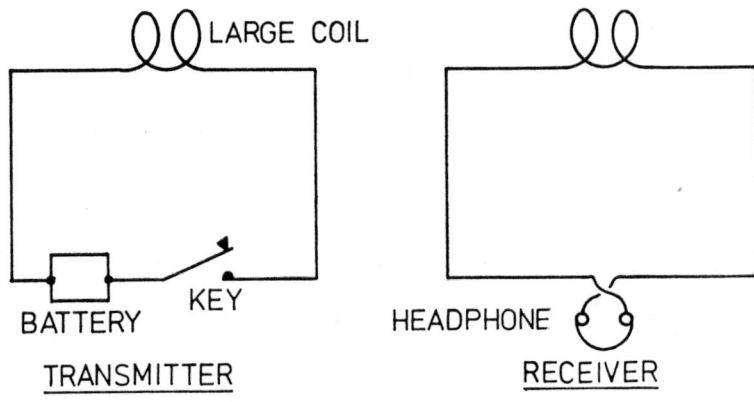

Fig 2 Induction system

was in fact the beginning of circuit tuning as we know it today and was one of the major discoveries leading to the development of practical wireless telegraphy.

It can now be seen that those scientists who were experimenting with the conductive and inductive systems of 'wireless' telegraphy were working on the wrong lines, though some of their discoveries, particularly Lodge's circuit tuning, were destined to form the basis of later developments.

Parallel with these experiments, others were taking place which were destined to be much more fruitful. All forms of radio communication as we know it today depend on electromagnetic waves radiated through the ether. How these waves were discovered is a remarkable story. Just as astronomers have calculated the existence and position of certain heavenly bodies before anyone has seen them, so the existence of electromagnetic waves was demonstrated theoretically before their existence had been established by practical experiment. In 1864 James Clerk Maxwell read a paper entitled 'A Dynamical Theory of Electro-Magnetic Field' before the Royal Society. In it he showed that electromagnetic waves must exist and that they could be propagated through space.

Twenty-three years later, in 1887, the German scientist Heinrich Adolph Hertz succeeded in producing these waves; his transmitter was based on the induction coil, an instrument used in laboratories for producing high voltages, and described in Appendix A (p 207).

The receiver which Hertz used to detect the waves consisted of a single metal loop, the ends of which were separated by a minute gap. The experiment was conducted in a darkened room, and when the transmitting spark occurred the tiny speck of light could be observed. The device became known as the Hertz resonator. The electromagnetic waves which made this experiment possible were for a long time called Hertzian waves. Hertz found that they travelled at the speed of light, 186,000 miles (3,000 million m) per second, and also that they could be reflected by metal objects, a property which Marconi tried to make use of in his early experiments; but this quality did not have a practical application until the development of the beam system.

The Hertz resonator fell a long way short of being a satisfactory detector but the first step towards producing something better had already been taken. In 1879 Professor D. E. Hughes, inventor of the printing telegraph, discovered that a tube of metal filings was sensitive to an electric spark near it; he was able to obtain a response from a tube of filings in circuit with a battery and telephone placed 500yd from the spark. This discovery led a few years later to the invention of the coherer described in Appendix A (p 215).

In his autobiography *Past Years* Sir Oliver Lodge describes how this instrument was invented in 1890 independently by himself and M. Branly of Paris. At the time Lodge was working on problems arising from Hertz's discoveries, work which later led to the development by his sons of the sparking plug which bears his name. A similar coherer was developed about the same time by the Russian scientist A. S. Popoff. Other investigators were also working along the same lines.

We now come to a man whose name is associated above all

others with the development of wireless telegraphy. Guglielmo Marconi was born in Italy in 1874, the son of Guiseppe Marconi, a landowner with an estate near Bologna, and his wife, Annie, an Irish Protestant. His mother insisted on teaching him English, and she brought him up in her own Anglican faith, which, in Catholic Italy, may be taken to show that the family was one of liberal views.

Marconi wanted to join the Italian Navy as an officer, but was not considered bright enough. A second choice was the University of Bologna, but again he was rejected. However, he was allowed to attend lectures given by the physicist, Professor Augusto Righi, who was at the time carrying out experiments with Hertz's oscillator. These captured Marconi's imagination to such an extent that he started experimenting on his own, and it was not long before he succeeded in making an oscillator placed at one end of a room ring a bell at the other. His mother encouraged him, but his father told him that he was wasting his time. And Professor Righi said he should get down to serious study instead of repeating experiments that everybody knew about. But Marconi saw the possibilities in what he was doing and knew that he had to realise them before someone else did.

During this period Marconi was spending some of his spare time reading to an old man who was nearly blind. This man had been a telegraphist and he taught Marconi the Morse code. He could hardly have made a more valuable return for the young man's kindness.

Undeterred by family and professorial discouragement Marconi advanced rapidly. The Hertz oscillator, as we have seen, radiated by means of two metal plates. Marconi's first original contribution to the technique of radio was to replace one plate with a sheet of foil hung from a pole and the other with a plate laid on the ground. This greatly increased the range. The next step was to replace the sheet of foil on the pole with a single elevated wire. This advance is claimed for Popoff, and it is not clear whether Marconi invented it independently. Marconi also greatly improved

Experimental

the coherer, making it operate a Morse inker (the instrument then used in line telegraphy to record the incoming messages in dots and dashes on a strip of paper). And he continued to cover longer and longer distances, until the day came when he succeeded in transmitting from his home to a receiver operated by his brother on the other side of a hill.

His father now changed his attitude, for it had become clear to him that his son's work had great possibilities. He announced that it was time to obtain official backing and after much consideration Marconi approached the Minister of Posts and Telegraphs. But the minister was not impressed and no help was forthcoming from that quarter. It was a severe blow but not a fatal one. If official Italy would not back him, he must look elsewhere. Degna Marconi, in *My Father Marconi*, tells us how Marconi's mother pointed out that his invention's greatest potential lay in communication with ships at sea. England was the greatest maritime nation and England was therefore the place to go. Marconi agreed. She wrote to a nephew of hers, Henry Jameson-Davis, a successful young engineer in London, who agreed to help, and mother and son crossed the Channel in February 1896.

Jameson-Davis gave Marconi a letter of introduction to Mr (later Sir) William H. Preece, engineer-in-chief to the Post Office. Preece had himself done a good deal of work on what was then called 'space telegraphy', mainly with the inductive system. He was very favourably impressed with Marconi and what he had to say and show him and he placed the facilities of the Engineering Department of the Post Office at his disposal.

This is how it came about that it was the British and not the Italian Post Office that backed the young inventor and that the headquarters of Marconi's commercial activities were established in London and not Rome or Milan.

In reality Marconi's success to date was still very limited, his apparatus being hardly more than a scientific toy. He now set about turning it into something of practical use. How he did this is described in Appendix A (p 209).

Throughout this period Marconi received unstinted assistance from Sir William Preece and the Post Office and with their cooperation he carried out a series of experiments from the island of Flatholm in the Bristol Channel. Then at the request of the Post Office he gave a demonstration of transmitting over open country. For this he erected a temporary station on Salisbury Plain. In view of developments some twenty years later, an interesting feature of this was that he placed reflectors roughly made of thin copper sheets on either side of the receiver, as he did when giving a demonstration of directional transmission on the roof of the General Post Office in July 1897. Representatives of the Post Office and of the services were present. The demonstration was judged a great success and received much publicity throughout the world.

In contrast to his early treatment by Italian officialdom, Marconi seems to have been given every encouragement in England. His achievements not only seized the imagination of the general public, very few of whom would have had any idea of the technology behind them, but he experienced very little opposition from those in high places who might have been expected not to welcome innovation. The principal scepticism he encountered came from certain scientists who declared that he would not be able to accomplish his ambition to span the Atlantic because the electromagnetic waves would not follow the curvature of the earth.

By 1897 Marconi felt that he needed the backing of a commercial company. Negotiations were opened, and as a result The Marconi Wireless and Signal Co Ltd was formed, with capital largely subscribed by Marconi's mother's family; it opened for business at 28 Mark Lane, London EC. One of its first and indeed historic achievements was the opening, at Alum Bay, near the Needles, Isle of Wight, of the first coast station in the world intended for communication with ships at sea. It was from this station that on 3 June of the following year Lord Kelvin sent the first paid radio message. This was transmitted to a second station recently opened at Poole, Dorset, and forwarded by land line to

Glasgow. The addressee, the first person to receive such a message, was Sir George Stokes.

Shortly after being opened, these two stations were used for a very important experiment, by which Marconi showed that the ether could be made to carry two messages at the same time. Using two pairs of receivers and transmitters tuned to different wavelengths he successfully transmitted two messages simultaneously from Alum Bay to Poole.

It was from Ireland that Marconi received his first opportunity to demonstrate the commercial value of radio. There was to be a regatta off Kingstown (now Dunloughaire), and the Dublin *Daily Express* commissioned him to transmit reports of the races from a vessel near the course to its office in Dublin. He installed instruments aboard the tug *Flying Huntress* (so making her the first vessel to be fitted with wireless telegraphy) and at the harbour-master's office ashore. During the regatta, held 20–22 June 1898, he transmitted race-by-race reports to the shore station and these were forwarded by telephone to Dublin and posted as they were received in the *Daily Express* window. The result was sensational.

This was the first commercial use of wireless and the first use of it for reporting a sporting event. It can, in fact, be said that in addition to all his other distinctions Marconi was the first radio sports commentator, giving the first running commentary of a sporting event.

This achievement was noted in the highest circles and barely a week had passed when Marconi received a royal command. The Prince of Wales, later King Edward VII, while attending a ball at the Rothschild mansion in Paris, fell and wrenched his leg badly. When he was sufficiently recovered to travel, he returned to England, but instead of recuperating under his mother's roof at Osborne House, Isle of Wight, he decided to stay aboard the royal yacht *Osborne* anchored in the Solent. This did not please Queen Victoria, who was anxious about him. She had heard of Marconi's success off Kingstown and decided to call in his assistance. Summoning him to Osborne House, she directed him to establish a wireless link with the royal yacht. He set up a temporary station in the grounds of Osborne House and installed equipment aboard the yacht, thus enabling the queen to receive 150 bulletins concerning her son's health, some of them of considerable length.

Later that year radio moved into the field with which it has been associated ever since—the safety of life at sea. Trinity House, the corporation responsible for maintaining the lighthouse service and other aids to navigation round the British coasts, requested the Marconi Company to establish communication between the South Foreland Lighthouse situated on the edge of the white cliffs north of Dover and the East Goodwin Lightship, stationed twelve miles away in the middle of the Straits of Dover. Apparatus was installed and the wireless link went into service on Christmas Eve 1898.

An interesting feature of the installation was the arrangement by which a button pressed aboard the lightship caused a bell in the lighthouse to ring. It was quite simple. The output from the receiving coherer instead of passing to the Morse inker was switched to a relay which closed a bell circuit as soon as any signal was received. This may be regarded as the first radio calling device. That it operated satisfactorily can be attributed to the fact that there was no interference; the ether soon became much too congested for such a primitive device to be of any use. Fourteen years later the need for a calling device was tragically demonstrated when the distress call of the sinking *Titanic* was unheard by a ship lying almost within sight, because her one operator had gone off duty. It was not until well into the 1920s that a calling device capable of passing the rigorous official tests was produced, and then it was a very complicated instrument.

It was not long before this new link dramatically demonstrated its value. In January 1899, only a month after it had been installed, heavy seas damaged the lightship's bulwarks, and this was reported by radio to Trinity House. And as if this were not enough, on 3 March 1899 the SS *R. P. Matthews* ran down the lightship. A message was sent to the South Foreland and lifeboats were sent to the lightship's assistance. This was the first occasion on which a vessel at sea summoned help by radio.

Later that year 'Tommy' Lipton was to make his first bid for the America's Cup in *Shamrock I*. The contest was to take place off the coast of Long Island, and the *New York Herald*, no doubt impressed by Marconi's success at the Kingstown regatta, invited him to cross the Atlantic and perform a similar service for it. Marconi accepted and repeated his success, his radio reports of the races proving a great sensation.

While in New York he took the opportunity to give a demonstration to the American Navy. He and his two assistants installed equipment aboard two warships, the battleship *Massachusetts* and the cruiser *New York*, and signals were exchanged up to a distance of thirty-five miles. Unfortunately, friction arose with the naval authorities. They asked Marconi how he was able to exclude interference between one station and another but Marconi refused to reveal details of his circuit as it had not yet been patented. However, friendly relations were restored and a number of American warships were fitted with Marconi installations.

Marconi and his assistants returned to England aboard the American liner *St Paul*. As soon as they were aboard they installed radio equipment and on arrival in the English Channel communication was established with the Alum Bay station at a distance of sixty-five miles. Alum Bay then transmitted to the *St Paul* a bulletin on the South African War, which was then at a critical stage. The bulletin included the following two items.

> 3.30—Ladysmith, Kimberley, and Mafeking holding out well. No big battle. 15,000 men recently landed.
>
> 3.40—At Ladysmith no more killed. Bombardment at Kimberley effects the destruction of one tin pot. It is felt that the period of anxiety and strain is over and that our turn has come.

This, the first ever radio news bulletin, was hurriedly printed and, under the title of *Transatlantic Times* with the date 15 November

1899, it was published and sold to the passengers at $1 a copy in aid of a seamen's charity. This was the first newspaper carrying news supplied by radio to be produced afloat.

On the eve of the twentieth century, wireless telegraphy had demonstrated its value in transmitting news instantaneously from an otherwise inaccessible point, and for receiving it for local dissemination; it had shown its capability in the accurate transmission of personal messages; most important of all, it had shown that it could summon aid to a ship in distress. Even with its still very limited range it had convincingly demonstrated its worth.

What impact had it had to date on the world at large? Actually very little. Both the British and American Navies were certainly taking a very serious interest, but the British Army, after trying it without much success in South Africa, was doing very little about it. One or two of the big shipping lines were showing signs of interest, but so far it could fairly be said that as yet it had not really changed anything. Its successes, sensational though they were to those who read about them in the press, did not actually affect more than a very limited number of people, in this way the situation resembling that created much more recently by the moon landings. People marvelled but did not comprehend. However, to those who could see ahead it had opened up a vast and exciting new field of progress.

At this point it seems relevant to ask: who invented it? Unquestionably the answer which many people would give to this question would be, Marconi. However this, like so many easy answers, needs to be closely examined. In previous pages we have mentioned the work and discoveries of numerous other investigators and there is no doubt that some of these have a claim to consideration. It could be suggested, for instance, that the honour should go to Hertz, the first man to propagate electromagnetic waves, but in fact he only forged one link in the chain of discoveries and inventions which started with Clerk Maxwell the mathematician and which still, more than a century later, shows no sign of approaching completion. Had Hertz not died at the early age of thirty-six, in 1894, the year in which Marconi started, he might very well have made further valuable contributions. As it was these were left to others.

Russia claims the honour for A. S. Popoff, but this cannot be accepted. Certainly he was a contributor, but no more. In order to arrive at an answer, let us look again at Figs 11 and Fig 12. These illustrate in diagrammatic form the transmitter and receiver used by Marconi in 1899. Neither of them was a single instrument, but each was an assembly of components developed by different investigators. Table 1 shows who was responsible for each part.

TABLE 1

Unit	Inventor	Nationality
OSCILLATOR, comprising induction coil, condenser and spark gap	Hertz	German
HF TRANSFORMER	Marconi	Italian
ELEVATED WIRE AERIAL	Marconi	Italian
EARTH CONNECTION	Marconi	Italian
HF TUNED RECEIVING CIRCUIT	Lodge	British
COHERER DETECTOR	Lodge	British
	Branly	French

The induction coil, the Leyden jar condenser and the Morse inker were standard instruments developed for other purposes.

We can see from the above that no one man can be named as the 'inventor', but that at least four investigators can fairly be said to have made a significant contribution to the fashioning of wireless telegraphy in its earliest practical form—five if we admit Popoff's claim to be the inventor of the elevated aerial. It is interesting to note what an international group they were, representing four countries, six if we include Popoff and remember that Marconi was half Irish. There were in addition a number of workers whose discoveries influenced the final result, though to a lesser extent, and there were others whose inventions had not yet been fully developed but which were destined to be of great value. Among these were Professor E. Rutherford, who did pioneer work on the magnetic detector from 1895 to 1897, and Nicola Tesla who invented the synchronous spark gap at about the same time. We have not mentioned the great American inventor, Dr Lee De Forest, but in fact he started work in this field rather later than did the others and his immensely important triode or three-electrode valve was not announced until 1907.

Whatever assessment we may make now, in the popular mind Marconi will always be the inventor of wireless telegraphy. However, it is clear that he was not entitled to all the credit, nor is there any evidence to suggest that he ever claimed it.

Marconi has also been described as the inventor of wireless signalling, but that too is rather doubtful as he was not the first man to see that Hertzian waves might be used for practical signalling. In an article published in the *Fortnightly Review* of 20 February 1892 Sir William Crookes suggested that it might be possible to use Hertzian waves for 'space telegraphy' and that this might become of international importance, but he did not suggest how it could be done. Lodge, also, suggested that it could be done, but he too did not follow the idea through to completion, though he did do a lot of work on it and he was the only man other than Marconi to come near to being popularly acclaimed as 'the inventor of wireless'. But he did not devote to it the undivided energy and attention which is essential to success in any great enterprise. He admits this in *Past Years*. Describing his return to work after a holiday in 1888, a year in which he had been working on the problems connected with the propagation of electromagnetic waves, he wrote:

> So home to the neglected waves, though I did not follow them up as I should have done. I left them to Hertz mainly, though I did devise a coherer method of receiving them. Lord Rayleigh, on congratulating me on this said, 'There's a life's work in it.' But I did not follow it up effectively.

At the time Lodge was forty-one and professor of physics at Liverpool University.

'Effectively' is the operative word, for Lodge did continue to experiment in the field of wireless, making some important contributions, and in 1911, when the Marconi Company bought the rights in the Lodge–Muirhead Syndicate's patents, he joined it as scientific adviser. Today he is remembered mainly for his work in other fields. Most of the other early pioneers, among them Rutherford and Edison, are remembered, if at all, like Lodge for other contributions to knowledge. Only Marconi is today generally remembered for his contributions in the field of wireless. How did this come about?

The reason is that he differed from the others in at least one important respect. It is probable that all the others would have registered very high IQs, if anyone in their day had known how to assess them, no doubt much higher than Marconi, who as we know failed to matriculate. But they were mostly academics, engaged in the pursuit of knowledge for its own sake, ivory tower men rather than potential tycoons. Marconi, on the other hand, though not cut out for academic distinction, had energy and drive, and a determination not to allow others to reap the rewards of his work. He saw from the first the possibilities inherent in what he was doing, though even he cannot have guessed in those early years just how great those possibilities were. He pursued his ambition with complete singleness of purpose, bringing together the contributions of others to build up a workable whole. Perhaps more than anything he was that somewhat rare bird, an inventor who was also a good businessman.

One thing is certain, Marconi stood head and shoulders above the many who had by then joined the radio gold rush. And that gold rush, of which he was at only twenty-five the leader, was to produce richness far in excess of what even the most imaginative of those engaged in it could have guessed at, and which was in fact man's first step on his journey to the moon.

2

Mushroom Growth

With the value and practicability of wireless telegraphy firmly established, the next step was to put it to work in the service of man. Its most obvious use was for communication with ships at sea, which had been in Marconi's mind from the first, and here, both in the navies and in the mercantile marines of the world, it had a clear field.

It is interesting to consider just how total was the lack of competition. At the turn of the century a smart liner carrying several hundred passengers, many of them people of importance, and a large crew, could sail out of port, drop her pilot, and disappear over the horizon; nothing would then be heard of her until she reached her destination, perhaps a week or even a month later. If she did not appear on the expected day, anxiety would soon build up among owners and relatives. And if in fact the ship never appeared, nobody would ever know what had happened to her; this is what actually occurred in the 1880s when the White Star liner *Pacific* disappeared without trace, and it happened again in 1909, as we shall see later.

While those ashore were cut off from the ship, her passengers and crew were equally out of touch with the course of events ashore. In a month of no contact with the outside world, who could tell what had blown up—a financial crisis, the fall of a government, perhaps even a war? An extreme example of how great this uncertainty could be was the dilemma in which Sir

Francis Drake found himself on returning from his historic circumnavigation. After three years away there was no knowing who was in power and who was in favour at court, and he had to approach his home port with extreme caution. On the domestic level, too, passengers and crew were unable to communicate with their families, which must have been particularly distressing in time of personal crisis; and the businessman was completely out of touch with market movements and with the conduct of his business.

The pioneers of wireless, faced with the huge task of filling this vacuum, had to start building an entirely new industry. The first company in this virgin field was, as we have already noted, Marconi's Wireless & Signal Co Ltd, which was soon to change its name to Marconi's Wireless Telegraph Co Ltd. It set about its task with vigour, beginning by launching, in quick succession, several subsidiaries, of which the most important were the Marconi International Marine Communication Co Ltd, which handled wireless aboard British merchant ships, and Marconi's Wireless Telegraph Company of America Inc, which later became one of the constituent companies forming the giant Radio Corporation of America (RCA). Others were formed to develop the company's business in Canada, Spain, Argentina, France, Germany and elsewhere.

There also came into existence a number of rival companies. In Great Britain the Lodge–Muirhead Syndicate was formed to exploit the patents of Sir Oliver Lodge and Dr Muirhead; it was to remain a competitor of the Marconi Company (as we shall refer to the Marconi group from now on) until it was absorbed by it in 1911. In America there was strong competition from De Forest's interests, and in Germany the Telefunken Co was formed to exploit the patents of Dr Slaby and Dr Wein. The Marconi Company, however, was in a very strong position with regard to patents, and the attempts of competitors to develop systems which did not infringe led to numerous patent suits.

However, the instruments used—the transmitter, based on a

24V battery and a spark coil, the receiver on the coherer—were still very primitive, with ranges of the order of sixty or seventy miles, sometimes a little more. Such distances were of very limited use in the open Atlantic and where ships were two or three hundred miles apart. Furthermore, the speed of transmission was limited to about ten words per minute, quite insufficient to cope with a large volume of traffic. Much remained to be done before wireless telegraphy could be regarded as a profitable commercial enterprise, which is what Marconi was determined it should become. From the first, every increase in its commercial and social impact was achieved by new invention; each technical advance was followed by an expansion in the fields already covered and every entry into a new field was the result of such advances.

Patent Office records show that by the turn of the century there was already a steady stream of inventions coming in relating to wireless telegraphy. By the mid-1900s that stream had become a torrent, as it has continued to this day. As is the case in all fields, many of the inventions never got far outside their authors' workshops or laboratories, and some of them look very odd indeed. But when these had been discarded there were several which were of great importance and which quickly put wireless on a sound basis. Some of these will be discussed briefly here. Fuller accounts of them will be found in Appendix A.

From about 1900 onwards there were two main streams of development in wireless telegraphy. One was related to its use for communication with mobile stations, at first mainly ships, the other with point-to-point working, usually over long distances.

The weak point of the early Marconi system was the receiver. The coherer was far from sensitive and it needed a good deal of skilled attention. Also it was slow in operation. Its output was fed into an ordinary telegraph Morse inker which printed the incoming signals on paper slip or tape in dots and dashes. With this arrangement it was impossible to separate the signal required from an interfering signal. An invention which greatly improved

Page 33 (*top*) An early Marconi coherer receiver; (*bottom*) an early Marconi wireless telegraph transmitter

Page 34 (*top*) The Poldhu transmitting apparatus. On the extreme left are the transformers; the banks of condensers are carried in metal containers in the wooden rack; on the right is the spark gap; (*bottom*) the earliest form of mobile radio (1901). This steam bus, by Thorneycrofts, was used in experiments with the Haven Hotel station. The cylindrical aerial could be lowered to a horizontal position when the bus was on the move. Marconi stands at the extreme right

reception was Lodge's replacement of the inker by a telephone earpiece, which was described in Patent No 11575 of 1897. The telephone was more sensitive than the inker, enabling the operator to distinguish between interfering signals, but it was not adopted generally for some years. This was probably due to a carry-over from line telegraphy, a preference for a receiver which delivered a record of the message on paper.

Between 1895 and 1897 Professor E. Rutherford had experimented with a receiver based on the discovery that electrical oscillations passed through a coil surrounding a magnetised iron wire will demagnetise the wire. Marconi developed this into a practical instrument, and in 1902 aboard the Italian warship *Carlo Alberto* he demonstrated his magnetic detector. It was an immediate success. It was sensitive, reliable, stable and robust. For many years it was a standard fitting in the majority of Marconi stations, and it was not finally superseded until the end of World War I. Equally important was the development of various forms of crystal detector invented by General Dunwoody in the United States, Professor G. W. Pierce, and others, which had an even longer life; and a third was R. A. Fessenden's electrolytic detector which was extensively used in Germany and the United States.

The design of transmitters was also greatly improved. The battery operated spark-coil was superseded by a transformer taking current from an alternator, thus enabling power to be very greatly increased. New types of spark gaps (described in Appendix A) were also introduced, giving further improved efficiency and selectivity.

The Marconi Company incorporated these improvements in its $1\frac{1}{2}$kW set introduced in 1905, which made wireless a sound commercial proposition. This set had a nominal daylight range of 250 miles, though much more could be obtained at night. With it an Atlantic liner could maintain communication with the shore during a large part of its voyage. The sets were extremely robust and electrical or mechanical failures were very rare. Later the fixed spark gap was replaced by a rotary gap, and with this

c

modification some of them were still in use well into the 1930s.

The first liner to be fitted with WT commercially was the Norddeutscher-Lloyd *Kaiser Wilhelm der Grosse*, equipped in 1900. In November of the same year the Belgian cross-Channel steamer *Princesse Clementine*, engaged in the Ostend–Dover service, was fitted. She communicated with a coast station at La Panne, near Ostend, and even with her primitive equipment she managed to remain in touch with the shore throughout her short voyages. From the point of view of publicity the Marconi Company was fortunate to have been given the contract, for she was soon able to demonstrate the practical value of wireless in an emergency. On 1 January 1901 she reported that the barge *Madora* of Stockholm was waterlogged and aground on the Ratel Bank and a tug was sent out to assist the barge into port. Only nine days later the *Princesse Clementine* herself ran aground in thick fog at Mariakerke and was able to summon help by wireless.

The *Lake Champlain* of the now defunct Beaver Line, when she sailed from Liverpool for Canadian ports on 21 May 1901, was the first British merchant ship to carry wireless. H. E. Hancock in *Wireless at Sea* describes what was in fact the first wireless-room aboard a British liner.

> As there was no accommodation available in the *Lake Champlain* for the wireless apparatus a room had to be built, and it may be of interest to compare it with the specially designed and equipped cabins of today. It consisted of little more than a cupboard 4 ft 6 ins in length and 3 ft 6 ins in width, one side being formed by the iron bulkhead. It was made of match-boarding without any windows, and when natural light was required the door had to be opened. The total cost of this structure was five pounds.
>
> The apparatus itself was mounted on a table covered with green baize, the accumulators being placed on the floor and the lamp resistance for charging the cells screwed to the wall. Two induction coils were provided, one of which was kept as spare. The two coil boxes, one on top of the other, served as a seat, the empty coil boxes providing a convenient cupboard for spares and sundries.
>
> F. S. Stacey, the first British radio officer, had a busy time

while the ship was in range first of Holyhead and then of Rosslare, handling a good many messages in each direction, but as the ship headed out into the Atlantic there was nothing more he could do, as there were no more British stations open and none on the other side of the Atlantic, nor were there any other ships with which he could communicate. Most of his time was spent demonstrating the apparatus to the passengers and, while the ship was in Halifax, to members of local scientific societies. On the return voyage he found that Crookhaven Station, in the extreme southwest of Ireland, had been opened, and he also established contact with the newly fitted Cunarder, *Lucania*.

To communicate with shipping the Marconi Company erected a chain of coast stations round the British Isles (see Fig 3). These were the three mentioned above plus sites at North Foreland, Caister-on-Sea on the coast of Norfolk, and Withernsea near Hull. Of these only North Foreland still survives. These stations and others erected later were made fully effective when in 1904 the Post Office undertook for the Marconi Company the collecting, transmission and delivery of all ship-to-shore and long-distance telegrams.

From 1900 onwards the erection of land stations spread rapidly. All over the world coast stations for marine communication and stations for point-to-point working sprang up. Parallel with this expansion there was a steady increase in the number of ships fitted. In contrast to the situation today, when practically every sea-going ship carries a wireless installation as a matter of course, in the early years of the century the wireless companies had to 'sell' the idea to the shipowners; and for many years they experienced considerable sales resistance. As we shall see, it was not considered necessary to fit even an important liner until as late as the autumn of 1908.

As one would expect, the first vessels to be equipped were mainly passenger ships, with the Cunard Line leading the way. To the liner companies wireless had a lot to offer. To begin with it gave prestige; in an extremely competitive market the phrase 'all

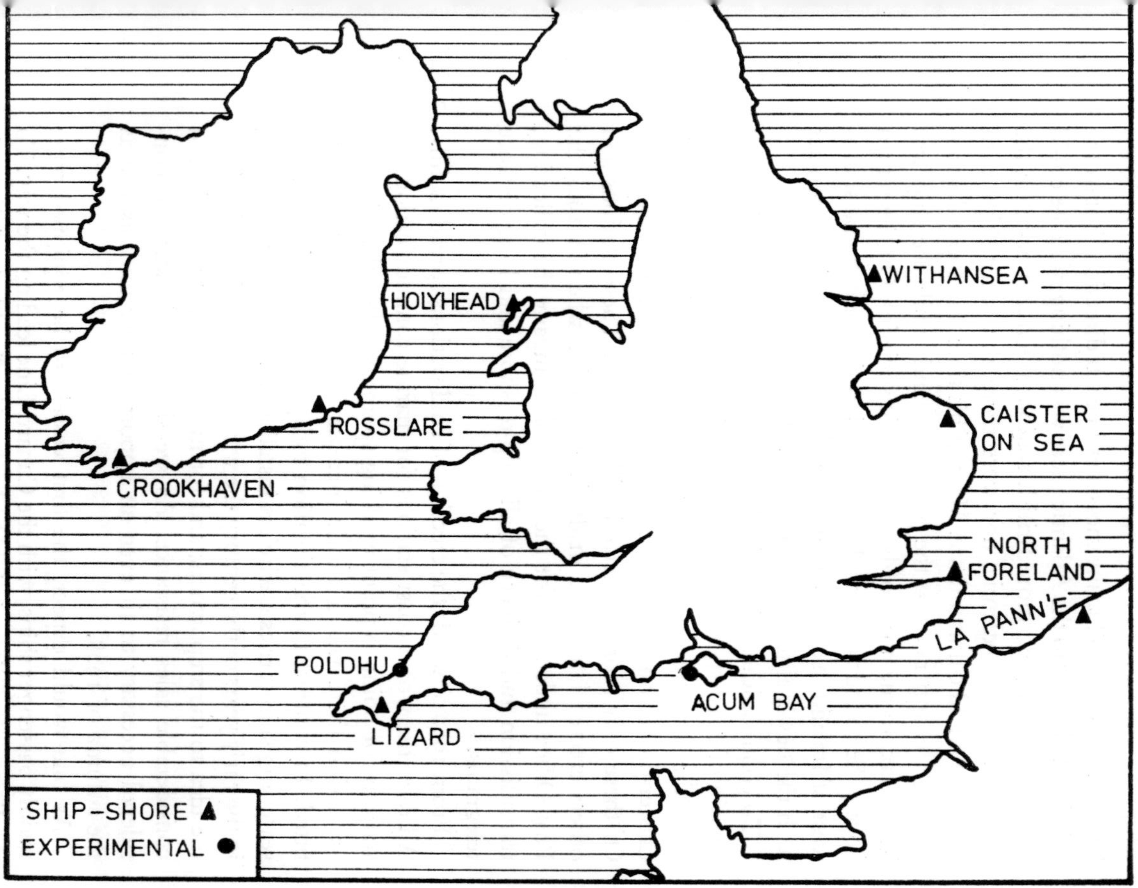

Fig. 1. First land stations, 1901

ships fitted with Wireless Telegraphy' was a strong selling point. And with ships running to very tight schedules, the facilities which wireless provided to give warning of expected late arrivals and other urgent matters relating to the movements of ships were very valuable. For passengers it was a great convenience to be able to notify friends of their time of arrival or to book hotel accommodation, and it enabled businessmen to keep in touch with the stock market throughout the voyage. Above all, there was the sense of safety which it conferred, and after the *Titanic* disaster wireless became a 'must' for passenger ships.

With cargo ships the position was different. For those on regular services the operational advantages which wireless conferred were considerable, but for the lesser ships—the coal and grain tramps, the Geordie colliers, the coasters—it was a different matter. A day or two's delay was often not important, whilst the cutting of costs most certainly was. And as for the safety of their crews, that was something which was not always of paramount importance to tramp-ship owners. Such ships were generally not fitted until the law came to require it.

The Marconi Company fitted and controlled the wireless stations on most British ships, and its subsidiaries obtained a large share of the market in fitting foreign ships. But rivalry, mainly from Germany and the United States, was not confined to that between the various sales organisations. To those familiar with the now strictly enforced regulations forbidding unnecessary and unauthorised transmission, the state of affairs in the ether of the 1900s must seem almost incredible, for this rivalry extended to the operators at sea, who deliberately jammed the transmissions of stations controlled by competing companies.

There were in fact no regulations regarding the use of the air, and there was nothing to stop anyone transmitting anything he liked and in any manner. That was all very well so long as there were only a few stations, and when ranges were usually less than a hundred miles, but as numbers and power rapidly increased it became clear that strict control was necessary if a state of complete chaos was to be avoided. This led to the first International Conference on Wireless Telegraphy, which was held in Berlin in 1903. At it and at subsequent conferences agreement was reached on regulations which enabled the wireless service to be conducted in an orderly manner by properly qualified staff and from stations licensed by the governments concerned. Great Britain ratified the convention and implemented it by passing the first Wireless Telegraphy Act, which came into force on 14 August 1904. It laid down regulations for the licensing of stations and the certification of operators, and it decreed that there must be no interference with British coast stations or naval working. It is noteworthy that only eighteen years after Hertz had succeeded in propagating electromagnetic waves for the first time, and during a period in which change was much less rapid that it is today, wireless telegraphy had grown to such an extent that it had made an international convention necessary.

The British government did not at that time make any regulations requiring ships to be fitted, but in the United States Congress passed an Act, which came into force in July 1911, prohibiting any ship carrying fifty persons or more from sailing from an American port if it was not fitted with an efficient wireless installation. As this applied to foreign vessels its effect extended beyond American-owned fleets.

However, the need to compel reluctant shipowners to have their ships fitted became increasingly apparent and international action was taken. One section of the International Convention for the Safety of Life at Sea, which came into force on 20 June 1914, laid down regulations requiring that various classes of ships, based on the number of persons carried, must be fitted with and maintain a wireless watch for not less than a prescribed number of hours per day. In July 1916 it became compulsory for all vessels of 3,000 tons gross or more to be fitted, and following the end of World War I this minimum was lowered to 1,600 tons.

The Marconi Company did not retain control of British coast stations for very long. On 29 September 1909 they were taken over by the Post Office, which was at the same time granted a licence to use all Marconi patents. Government control of coast stations soon spread and before long it became general throughout most of the world, though the United States retained the system of private control.

Today marine wireless provides not only a means of instant communication ship-to-ship and ship-to-shore, but also a number of ancillary services. These include, for the captain, time signals, weather reports, storm warnings, navigational warnings, direction-finding, radar; for the passengers, news and broadcast programmes from the shore. But in the early days of the century few of these services were available.

The first to be provided was a news service. We have seen that as far back as November 1899 the first news bulletin to be received aboard ship was published aboard the *St Paul*, but this was a once-only experiment. The first regular news service was opened by the Marconi Company in 1904, with nightly transmissions of news bulletins from Poldhu, Cornwall and Cape Cod, Mass. It was intended that these two transmissions should cover the whole of the North Atlantic route, but there was an intermediate area where it was often difficult to pick up either. Nevertheless, the operators on the transatlantic liners made it a point of honour to provide a daily news bulletin for the passengers. In those pre-valve days it was often necessary to pick up signals so weak that the slightest unwanted sound—even the operator's own breathing —was enough to drown them. Some operators went to considerable lengths to exclude extraneous noise, and old hands tell of one who had himself shut in the silence cabinet, a very small compartment encased in 6in of cork insulation intended to deaden the deafening noise of the spark; the hour he spent there copying out the news bulletin must have been extremely uncomfortable.

The news was printed on board and published in the form of a ship's newspaper. The first of these was the *Cunard Daily Bulletin* of which number 1 appeared aboard the *Lucania*. This was soon joined by others and by 1913 there were *The Ocean Times*, *The Wireless Mail*, *The Atlantic Daily News*, *Das Atlantische Tageblatt*, *Le Journal d'Atlantique*, *Giornale dell Atlantica* and *Diaro del Atlantico*. An important part of the bulletins was the stock market reports.

Today the whole of the western world and much of the rest depends for accurate time-keeping on frequent radio timechecks. On land watches are set by the familiar pips, and navigators, who depend on accurate time-keeping for determining longitude, can obtain at least one daily check wherever they may be.

At the turn of the century the situation was quite different. The Post Office and the railways sent daily time-checks to their local offices and stations, and the ordinary citizen had to check his watch or clock either from these or at second-hand from church or other public clocks; whilst the factory hooter or whistle set the time for workers to clock in. Navigators, who required accuracy more than anyone else, could obtain a check when in port from a 'time ball', which was a ball dropped from the gaff of a flagstaff on the harbourmaster's office or coastguard station; at sea they could obtain no check at all. On the global scale each country based its time on its own observatory's astronomical observations, but there was no means of checking these against each other.

It was the need for such international checks that led to the introduction of wireless time signals. In 1910 the first step was taken by the French Bureau des Longitudes which arranged with the Paris observatory to send out a time signal every night from the high-power military wireless station on the Eiffel Tower. Shortly afterwards a similar service was started from the German station at Norddeich. Navigators aboard ships fitted to receive them soon found these time signals to be of immense value, and in 1912 a conference was called in Paris to plan a worldwide time signal service. This drew up a list of stations (Table 2) which were to transmit time signals regularly.

TABLE 2

Station	Country
Paris, Eiffel Tower	France
Norddeich	Germany
Arlington, DC	United States
San Francisco	United States
Sao Fernando de Norohna	Brazil
Mogadisco	Italian Somaliland
Manilla	Philippines
Timbuctoo	French West Africa

The first four, together with some not on the list, were in operation by 1913, but it is doubtful whether all the others were working before the war broke out in 1914. Even if they had been, the whole eight would have given a great deal less than world coverage.

It is amazing that the list contains no station in any part of the then far-flung British Empire, nor was any such station projected. Considering that Great Britain was the power which would certainly have benefitted most from such a system, one would have expected that she would have taken the lead in establishing it. Even if it had been thought unnecessary to duplicate the comparatively nearby Eiffel Tower service, that would not excuse the absence of any service from such points as Cape Town, Singapore or Hong Kong. And indeed the matter did not escape attention at the time.

That there was an urgent need for more accurate navigation was demonstrated by the *Titanic* disaster. The *Carpathia* found her wreckage thirty-four miles from the position which she had given in her distress message, a discrepancy which may have been due to an error on either ship or to a combination of both; certainly it may well have been due to errors in time-keeping.

Accurate determination of longitude is essential not only in navigation but also for map-making. As soon as time signals were available they proved to be invaluable for this purpose in inadequately mapped areas. They were first used for map-making in 1913 in connection with the determination of the Brazil/Bolivia border, when the relative longitudes of Manaos in the Amazon Forest and Porto Velo were calculated. Shortly afterwards the Italian explorer Dr Phillipo de Phillipo, working in central Asia, used time signals from Lahore for a similar purpose.

Nowadays we associate radio and television very closely with the weather, and we expect to be told two or three times a day what weather we may expect later in the day, tomorrow, or for the month ahead. Accurate forecasts, instantly available, are considered indispensable in the fields of agriculture, shipping, and particularly aviation, and a large part of the information on which the forecasts are based is transmitted by wireless. What services were available before World War I?

The first attempt to make use of wireless for meteorological reports was made by the Weather Bureau of the US Department of Agriculture, when in 1900 it appointed R. A. Fessenden to investigate the possibility of distributing weather information by wireless. In Great Britain the use of wireless in the collection of meteorological information began in 1904, when messages containing such information were sent to the *Daily Telegraph* from ships at sea, but it was not until several years later that any official service was organised. A body called The Commission for Weather Technology was eventually set up which discussed the matter in 1909 and 1912. As a result, arrangements were made for twice-daily messages to be sent from both naval and merchant ships in the North Atlantic and the North Pacific to certain meteorological stations, and by 1913 a substantial number of ships were engaged in the service. Unfortunately, with the limited relay facilities then available, many of their messages did not arrive in time to be of maximum value. According to the *Yearbook of Wireless Telegraphy for 1913*, of a total of 4,990 messages sent to the Meteorological Office in London from naval and merchant ships only 232 arrived in time to be included in the

Daily Weather Report and 2,352 for inclusion in Yesterday's Weather Map. The weather reports compiled by the Meteorological Office were communicated to the press and can hardly have been of much use by the time they reached their readers.

The regular broadcast to shipping of weather reports and forecasts did not start much before 1912, but by the end of 1913 a service was provided by Great Britain, Australia, Holland, South Africa and the United States.

From April 1912 the United States Weather Bureau collected weather information by wireless from ships in the Caribbean and the Gulf of Mexico, and this was used to give warning of the approach of hurricanes.

The incursion of wireless telegraphy into the fields of time-keeping, navigation and meteorology, though of great importance, did not make headline news. Very different was its first use in the fight against crime.

The story of the murderer, Dr Crippen, is well known and all we need here is an account of the part played in it by wireless telegraphy. When Crippen found that London was getting too hot for him he decided to flee with Miss Le Neave to Canada. The route he chose was unorthodox. He first crossed to Rotterdam and there booked passages on the Canadian Pacific liner *Montrose*, sailing from Antwerp on 20 July 1910. Soon after the pair left London a warrant was issued for their arrest, and newspapers containing their description reached Antwerp before the *Montrose* sailed. Her commander, Captain Randall, read these and soon after sailing felt sure that he had spotted the two fugitives among his passengers. He sent a radiogram to Scotland Yard with the result that Chief Inspector Drew and two other police officers were despatched to Canada by a faster ship, the White Star liner *Laurentic*. Captain Randall also sent a radiogram of more than 1,000 words to the *Daily Mail*, which had offered a reward of £100 for information which would lead to an arrest. During the crossing the two vessels kept the press informed of their positions, and the news was presented as a race, followed by many thousands of readers. It was a race inevitably won by the faster *Laurentic*, which reached Canada in time to enable Inspector Drew to board the *Montrose* at Father Point in the St Lawrence estuary and arrest the fugitives.*

Another valuable service contributed to shipping by wireless telegraphy was the establishment, consequent on the *Titanic* disaster, of the International Ice Patrol. On 3 March 1913 the *Scotia*, fitted with a $1\frac{1}{2}$kW transmitter, sailed from Dundee to take up her position in the North Atlantic ice area, where she collected information on sightings of ice and broadcast it for the benefit of navigators.

Even before wireless telegraphy had established itself firmly aboard ocean-going ships, it showed signs of making itself felt in other fields. As early as 1900 the Metropolitan Fire Brigade showed an interest in it, and had equipment installed in its stations at Mitcham and Streatham, then both on the outskirts of London. In the same year Lloyds had several of its stations fitted.

Later, the fishing industry also began to show interest, and in 1913 experiments were carried out with a view to determining whether wireless would be any help to the fishing fleets. The trawler *Othello* and the carrier *Caesar* were fitted as an experiment during their operations in the North Sea, with satisfactory results. In the same year the Lackawana Railroad in the United States experimented with wireless telegraphy on trains, and it was not long before here too wireless was put to good use. The conductor of a long-distance train was taken ill, and in answer to a wireless telegraphy message both an ambulance and a relief were waiting when the train stopped farther down the line. Three years later, in 1916, a severe blizzard damaged telegraph lines and wireless telegraphy was used to control the trains.

* Captain Randall was to have another demonstration of the value of wireless three years later. He was in command of the *Empress of Ireland* when she was sunk in collision, also near Father Point.

Nowadays exploring and mountaineering expeditions carry radio equipment as a matter of course in order to keep in touch with base. It was first used for this purpose in 1913 by the Mawson Antarctic Expedition, which erected a station on Macquarie Island to enable it to maintain contact with the Australian mainland. It was also similarly used in the same year by the Massey-Baker Expedition in Papua. The Shackleton Antarctic Expedition was not so equipped, with the result that Shackleton had to make an appalling journey across dangerous seas in order to summon help.

Radio telemetry, or the control of distant mechanisms by radio, may be said to have had its beginnings in 1898, when the pressing of a button aboard the East Goodwin Lightship rang a bell in the South Foreland Lighthouse. By 1914 something a good deal more sophisticated had become possible, and the Marconi Company began producing apparatus for controlling fog signals on sites difficult of access, such as unmanned lights on dangerous rocks. A writer in the *Wireless World* reporting this described the feat as 'the extreme limit of difficulty'. Clearly he was unable to see into the future!

Later in the same year a torpedo-boat, the *Nautilus*, the engines and steering of which were controlled by wireless, was built privately in the United States and the American government became interested in it. But nothing seems to have come of it, for we do not read of any wireless-controlled warships being used in World War I.

The most important new development to take place during the latter part of this period was the direction finder. This may be defined as an instrument for determining the direction from which wireless signals are coming. De Forest and others had experimented with it in the early 1900s, using simple apparatus, and in 1906 two Italians, Dr Ettore Bellini and Captain Tosi, produced a system which still bears their names and which is in principle the same as that used today. Development was at first slow because the receivers of that date were not really sensitive enough; it was put into service for the first time by the French Department of Posts and Telegraphs, which erected an installation for the assistance of shipping at its Boulogne-sur-Mer station.

The rights in the Bellini-Tosi system were bought by the Marconi Company in 1912 and from that point development went ahead rapidly. The first ship to be fitted was the *Mauretania*, aboard which successful experiments were carried out. A crystal detector was still being used and the usual range limit was about fifteen miles, though occasionally much greater distances were achieved. Other early ships to be fitted were the *Eskimo*, engaged on the Hull–Christiana (Oslo) run, and the *Royal George*. This instrument was, like radar nearly thirty years later, ready just in time to play a vital part in the defence of Britain.

3
For Those in Peril

In Chapter 2 we described some of the benefits conferred on shipping by the development of wireless telegraphy during the years up to World War I, but so far we have touched only briefly on what was by far the greatest of these benefits, the ability it gave to a ship in distress to summon help.

From the earliest days of regulated wireless telegraphy a distress call has had absolute priority over all other traffic; when a distress call is heard all stations operating on the distress wavelength and not concerned with the call must remain silent. At first distress calls were prefixed with the letters CQ, meaning 'all stations', but as that prefix was also used for other general calls it was not sufficiently distinctive, and in 1904 the Marconi Company directed that all ships controlled by it should use CQD. Numerous explanations of what these letters stood for were soon in circulation among the lay public, among these 'Come Quick Distress' and 'Come Quick Danger' being the most popular. Others were more fanciful.

The letters were in fact made up of CQ, 'all stations', and D, 'urgent'. (These abbreviations, together with numerous others, were adopted from those in use in the British Post Office telegraph service, and became international.) The use of CQD was ratified by the 1908 Berlin Convention, but it was short-lived. With the rapid growth in the number of ships fitted, CQ was heard many times a day being used as a general call for various purposes, and this took from CQD the extreme urgency essential for a distress signal. Accordingly the London Convention, which came into force in June 1912, replaced it with SOS. The date is significant, as we shall see.

Again popular imagination gave meaning to the letters, with 'Save Our Ship', 'Sink Or Swim', or even 'Save Our Souls'. They do not, in fact, stand for anything; they are not even transmitted as three separate letters but as a single signal composed of three dots, three dashes, three dots. It was adopted because it stands out clearly from a jumble of other signals and so is easy to pick up. The distress call used in radio telephony, 'Mayday', is a corruption of the French *M'aidez*, 'Help me'.

The first distress call was, as we have noted, sent out by the East Goodwin Lightship early in 1899, though we have no record of what form it took. After this, wireless telegraphy was used by ships to summon help with increasing frequency, including in 1902 the destroyer *Recruit*, aground, and in 1903 the Red Star liner *Kroonland*, disabled 130 miles south-west of the Fastnet.

These and other incidents were impressive, but the value of wireless for saving life at sea was probably first brought home to the public in a really spectacular way in 1909. On 22 January the White Star liner *Republic* was sailing homeward bound from New York to Liverpool, when twenty-six miles south-west of the Nantucket Lightship, and in thick fog, she was rammed by the Italian ship *Florida*. Though badly holed and with the wireless-room damaged, the emergency transmitter was still workable, and Wireless Officer John R. Binns was able to send out a CQD call which was picked up and re-broadcast by the coast station at Siasconsett. Several ships heard it and headed for the *Republic*'s position, among them the White Star liner *Baltic*, which was the first to arrive on the scene. The *Florida* had no wireless, so it was impossible for the *Republic* to communicate with her as the ships drifted apart in the fog. However, before the *Republic* sank, all her passengers and crew were transferred to the *Baltic* and *Florida* without a single casualty. Seventeen hundred lives were saved.

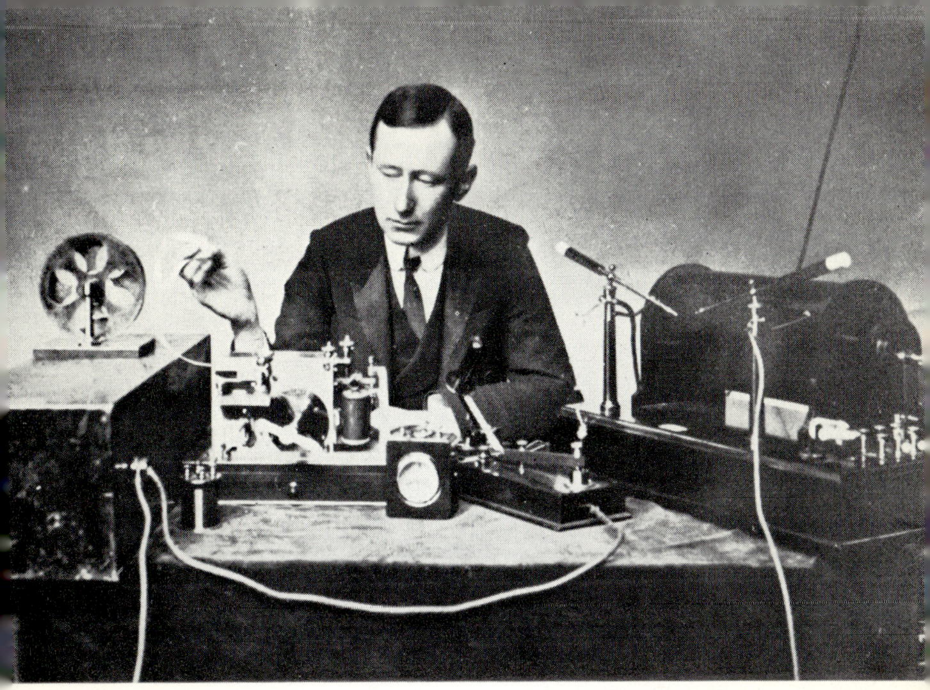

Page 51 (*top*) A still from the first cinematographic film about wireless, taken about 1901. It shows Marconi with typical apparatus, including a 10in induction coil spark transmitter (right), a morse inker, and a 'grasshopper' key in the centre; (*bottom*) King George V and Queen Mary beside a mobile wireless vehicle during a visit to the British Army on manoeuvres in 1912

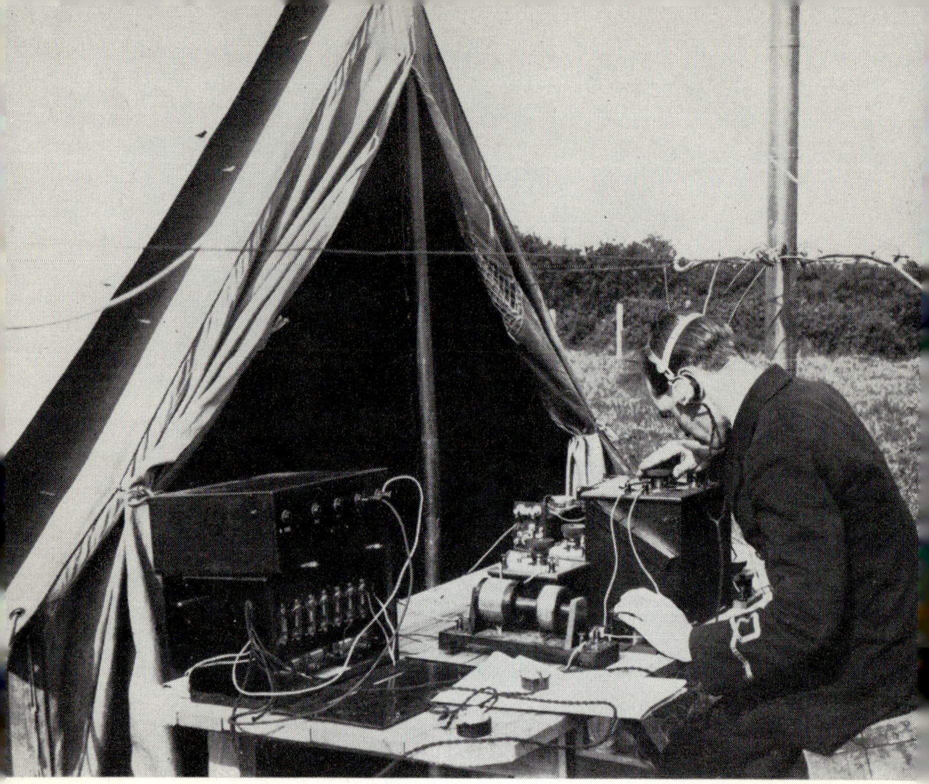

Page 52 (*top*) A small portable DF station (about 1913); (*bottom*) a bicycle power unit of World War I

This great achievement by wireless telegraphy created a tremendous sensation, and Binns received an offer to tour 'the halls' to demonstrate how he sent out the distress signal. But Binns was destined for a much more lucrative career, finally retiring as vice-president of the Hazeltine Electrics Co of New York.

This was not the only occurrence of this kind in 1909. In June the *Slavonia* was stranded in the Azores and again help was summoned by wireless; 410 lives were saved. But by far the greatest disaster to a single ship in this period was the sinking of the White Star Line's brand new *Titanic*, the largest ship afloat. The story of this calamity has been told so many times that there is no need to give a long account of it here. A study of the literature on the subject shows that the tragedy was the culmination of an almost incredible series of failures and wrong decisions, some of which are still, more than half a century later, subjects for controversy. Of one aspect of the disaster, however, there was never any doubt: from first to last the wireless service, both as to personnel and to the installations used, came up to the highest possible standard.

The liner struck the iceberg at 2340 hrs ship time on 14 April 1912, and at 0015 hrs on 15 April her commander, Captain Smith, ordered the senior wireless officer, J. G. Phillips, to send out a distress call. Phillips and his junior, Harold Bride, actually treated this order with a good deal of amusement. The White Star Line had so forcefully publicised their new ship as unsinkable that many of those on board refused to believe that there was any danger; when Phillips sent out the message, using the prefix CQD, Bride jokingly suggested that he should repeat it, using the new distress call SOS, which was not due to come into force for another two months, 'as he might not have another opportunity of using it'.

Some fifty miles away the small Cunarder, *Carpathia*, was proceeding cautiously on her passage from New York to the Mediterranean. She carried only one wireless officer, Harold T. Cotton. He had already gone off watch, but at the critical moment he decided to call MGY, the *Titanic*, on a routine matter before turning in. As he put the phones on he heard MGY calling CQD. The *Carpathia*'s commander, Captain R. H. Rostron, immediately altered course, and with a double watch of firemen shovelling coal into the furnaces, drove his ship at forced full speed through the pitch-black night and the icefield towards the position of the *Titanic*.

The sequence of events is recorded dramatically in the following extract from the *Carpathia*'s wireless log as quoted by H. E. Hancock in *Wireless at Sea*.

Sunday, April 14th, 1912.
N.Y.T.

5.10 p.m.	TRs [preliminary service messages] with *Titanic* bound west.
5.30 p.m.	Signals exchanged with *Titanic* at frequent intervals until 9.45 p.m.
11.20 p.m.	Heard *Titanic* calling SOS and CQD. Answer him immediately. *Titanic* says 'Struck iceberg come to our assistance at once. Position 41.46 N. long 50.14 E.' Informed bridge at once.
11.30 p.m.	Course altered; proceeding to the scene of the disaster.
11.45 p.m.	*Olympic* working *Titanic*. *Titanic* says weather is clear and calm. Engine room is getting flooded.

Monday, April 15th, 1912.

12.10 a.m.	*Titanic* calling CQD. His power appears to be greatly reduced.
12.20 a.m.	*Titanic* apparently adjusting spark gap. He is sending Vs. Signals very broken.
12.25 a.m.	Called *Titanic*. No response.
12.30 a.m.	Continue to call *Titanic* at frequent intervals but without success.

At daybreak *Carpathia* arrives on scene of the disaster.

Monday, April 15th, 1912 (continued).

5.5 a.m.	Exchange signals with *Baltic* but unable to read owing to continual atmospheric disturbances, etc.
6.45 a.m.	Inform him we are now rescuing *Titanic*'s passengers.

7.7 a.m.	Received the following message from *Baltic* to captain: 'Can I be of assistance as regards taking some of the passengers from you? Will be in the position about four-thirty. Let me know if you alter your position. Commander *Baltic*.'
7.10 a.m.	Sent following reply to *Baltic*: 'Am proceeding to Halifax or New York at full speed. You had better proceed to Liverpool. Have about 800 passengers aboard.'
7.40 a.m.	Advise *Mount Temple* to return to his course as there is now no need for him to stand by. Nothing more could be done. We have rescued twenty boatloads of the *Titanic*'s passengers.

Cotton does not seem to have heard the ice-warning sent out by the *Californian*, though unfortunately his times of going on and off watch have been omitted from the extract and he may have been off at the time. The difference in his times with those quoted earlier is due to the fact that he was keeping New York time. Apparently the universal use of GMT in marine radio had not then been introduced.

The sensation caused by the catastrophe exceeded that of any other news ever broken, except the declaration of a major war, and more than anything else it was the part played by wireless telegraphy that caught the imagination of the public. By 1912 wireless was no longer, strictly speaking, new. Most important ships including all British warships carried it, and a full-scale transatlantic service was in operation. But contact with it was confined mainly to the seafaring community, the more affluent travelling public and amateur experimenters. The *Republic* affair had been a nine-day wonder, lacking many of the sensational features of the sinking of the *Titanic*. With this great disaster the uses of wireless were inescapably brought to the attention of everyone who could read a newspaper. For the press it was a heaven-sent story, as the following extract from the London *Times*, dated 16 April 1912, shows.

Imagination is filled once more by the wonderful part played by wireless telegraphy in the story of the *Titanic*. The wounded monster's distress sounded through the latitudes and longitudes of the Atlantic, and from all sides her sisters great and small hastened to her succour. But for this new instrument of communication it might have been that the greatest product of naval architecture might have passed from our human ken, her fate for ever unknown. We recognise with a sense near to awe that we have been almost witnesses of a great ship in her death agonies.

In America reaction was very different, largely because of the unorthodox way in which the news reached New York. It is an interesting story which still contains an element of unsolved mystery.

At the time the two Wanamaker stores in New York and Philadelphia were linked by a pair of wireless stations. In charge of the New York station was a remarkable young man named David Sarnoff, who had entered the service of the Marconi's Wireless Co of America a few years earlier as a messenger boy and was eventually to become president of RCA and a leading figure in the development of broadcasting in the United States. While listening on the night of 14–15 April he picked up a message from the *Olympic* to the effect that the *Titanic* had sunk. This he communicated to the press in time for the morning editions. The effect of the news was staggering. Many people in the country had relatives and friends aboard, and the offices of the White Star Line and the newspapers were besieged by people clamouring for news of their safety, news which was not forthcoming.

Then, some time during the day, a message was received by Congressman J. C. Hughes which read as follows: '*Titanic* proceeding to Halifax. Passengers will probably land there. All well. —White Star Line.' This too was communicated to the press. At the same time numerous people possessing wireless receivers were listening in, trying to pick up news and passing on what they heard, and it may safely be assumed that this added to the confusion. In the afternoon the New York *Evening Sun* appeared carrying the banner headline ALL SAVED FROM TITANIC, stating that the passengers had been transferred to other ships and that the disabled liner was being towed to Halifax. At last an official announcement was posted outside the White Star Line's New York office, announcing the awful truth. Immediately there was a loud outcry by the newspapers against the wireless service for so long withholding the facts, although it was really they who were at fault for printing sensational but unconfirmed rumours as news.

All this time David Sarnoff remained on duty, as he appears to have done almost continuously until the *Carpathia* reached New York. According to Eric Barnouv in *A History of Broadcasting in the United States*, and others, he was the only source of news in the United States until the *Carpathia* docked. This can hardly be correct, for the *Carpathia*'s wireless log shows that Cotton despatched 157 messages during this period, among which must have been official messages from the captain to the Cunard and White Star offices in New York. Furthermore, the London press got the story officially, so it is not clear why the New York papers did not also get some authentic account from the agencies.

A study of Cotton's evidence before the US Congressional Inquiry goes a long way towards explaining how it was that the New York papers were unable to get in touch with the *Carpathia*. In the first place he stated that he refused to accept press messages for transmission, because he considered that messages from the survivors to their relatives were more important. Secondly, he stated that USS *Chester*'s attempts to communicate with him caused a lot of interference, because the operators on that ship were accustomed to using the American Morse Code and their efforts to use the International Code were unreadable. (The two codes are quite different. During the early years of WT on the North Atlantic US ships usually used the American code between each other.) Another difficulty arose at one point over the transmission of a long list of survivors, due to his own exhaustion after thirty-six hours of continuous duty.

The commissioners were very anxious to trace the origin of the message received by J. C. Hughes, and they closely questioned Cotton about it. In reply he was firm in his denial of ever having sent out a message stating that the *Titanic* was proceeding to Halifax. Who did send that message, and why, remains a mystery.

This catastrophe, which combined the destruction of what can fairly be described as man's greatest feat of construction to date with the terrible loss of life cannot, as we see now, be ascribed to a single blunder but to a whole sequence of follies, failures and mischances. Not the least of these concerned the *Californian* and her wireless station. This small Leyland liner was actually lying only ten miles away during the whole period when the *Titanic* was sinking, motionless because she was surrounded by ice and her commander, Captain Lord, was taking no chances. Her officer of the watch actually saw the *Titanic*'s rockets but (for reasons which we do not need to discuss here) did not call the captain, who had turned in.

The *Californian* carried wireless but had only one wireless officer; according to several accounts, at 2300 hrs on 12 April he started to send out a warning of ice but was told by Phillips on the *Titanic* to 'shut up', as he had traffic for Cape Race, Newfoundland. This sounds incredible. In the first place a navigational warning would have had priority over private telegrams. Secondly, it seems strange that the senior wireless officer on an important liner would have been so lacking in a sense of responsibility as to ignore such a warning. And thirdly, once the *Californian* had started transmitting, he would not have been able to hear any interruption.

It is all very strange. One thing, though, is certain: if the *Californian* did send out such a warning and it had been picked up by the *Titanic* the disaster might have been averted—provided it had been acted upon by Captain Smith, something that is by no means certain.

And how was it that the wireless officer on the *Californian* did not receive the distress call? For the simple reason that he had, quite properly, gone off watch and there was nobody to relieve him. It had already been recognised that wireless made every ship that carried it a potential lifeboat, but it was now demonstrated that the potentiality was reduced by something like 50 per cent if only one man was carried. However, on small ships it has never been possible, for economic reasons, to carry more than one, except under war conditions. If only, then, the *Californian* had been fitted with a device which could call an officer off watch, the officer would have been alerted. Well, why not? This was a question that was asked at the inquiry. Quite simply the answer was that radio technology had not yet reached such a point. Commandatore Marconi, as he had now become, assured the court that such a device would be available within the next year or two, and he suggested that it might be done with a 30 second dash substituted for the existing distress call. He was unduly optimistic. It was indeed possible to construct a device that would respond to such a dash and ring a bell, but unfortunately it would respond to a lot else. A large proportion of false calls would soon destroy confidence in it, and no notice would be taken of the bell. In the event it took not two but fifteen years to produce a device capable of passing the rigorous official tests.

We have dealt with the part played by wireless telegraphy in the story of the *Titanic* tragedy at some length because it marks a turning-point in the history of radio communication. There can be no question but that the value of wireless telegraphy for saving life at sea had already been conclusively demonstrated on numerous less spectacular occasions but the loss of the *Titanic* was so sensational that it brought home to everyone the existence and potential of wireless.

One phrase in the above quotation from *The Times* has special significance: 'Her fate forever unknown'. Of course before wireless telegraphy became available countless ships had sailed out of port never to be seen or heard of again. But even in the days of radio communication this was something that was still happening;

not quite three years earlier, in July 1909, a sizable passenger ship not fitted with wireless had disappeared, her fate to this day unknown. Wireless telegraphy was still regarded as an optional extra not only by tramp owners but even by some liner companies. Among these was the now long defunct Blue Anchor Line, the disappearance of whose *Waratah* still remains one of the great mysteries of the sea.

The *Waratah* was built in Scotland and completed in the autumn of 1908. According to Bryan Reed's *Great Mysteries of the Sea*, her gross tonnage was 9,333, comparatively large for the Australian trade at that time, and her speed was 13 knots. She was regarded as a well-found ship and was licensed to carry both ordinary passengers and emigrants; in fact an important addition to the fleet which her owners operated for their service between the United Kingdom and Australia. For all that, she carried no wireless.

On 26 July 1909, homeward bound from Australia for the second time, she sailed from Durban for Cape Town with 211 passengers on board, but never arrived. Soon after she left Durban the weather round the South African coast deteriorated. She was sighted once, perhaps twice, but after that she was never seen again. As to how she—a well-found, new ship—met her fate not one shred of positive evidence has ever come to light. There have been many theories of course. She might have been struck by a giant wave and capsized. She might have had an engine or steering failure and drifted helpless into the trackless, ice-strewn waters of the Antarctic. If she had carried wireless and no distress call was heard, the first alternative would have been indicated. If she had been disabled, she would have been well within range of help. But her owners apparently did not think wireless to be worth the cost, so we shall never know what actually happened.

Not much more than a year after the *Titanic* sank there was another great maritime disaster in which wireless telegraphy again played a vital part. On 13 May 1913 the Canadian Pacific liner *Empress of Ireland* with 1,407 people on board was steaming up the Gulf of St Lawrence, bound for Montreal, when she was in collision with the Norwegian freighter *Storstad* and very badly holed. The conditions were quite different from those that had obtained with the *Republic* and the *Titanic*, on both of which occasions there had been ample time to summon help. This time it was immediately apparent that the ship was going down fast. First Wireless Officer Harold Ferguson immediately sent out an sos, which was picked up by the coast station at Father Point. Then, only eight minutes after the impact, the power supply failed. Two rescue vessels were immediately despatched to aid the stricken liner but arrived too late to save more than a small fraction of her passengers from perishing in the icy water. There were only 444 survivors, but it is safe to say that there would have been very few indeed if it had not been for that sos message.

In August of the following year World War I broke out and wireless calls for help ceased to have any novelty. Now, frequently, the operators concerned had considerably less time to do the job than had Harold Ferguson. With a large hole torn in her side by a torpedo or a mine a ship could go down in two or three minutes, and if she was so unfortunate as to be loaded with iron ore, even less. To cope with such situations a very compressed form of distress message was devised. The following is an example, with its interpretation:

SOS SOS SOS	
de	from
ABC ABC ABC	Ship's call sign, indicating her name
SSS	I am being attacked by a submarine
5120 0721 nw	My position is latitude 51° 21′ N, longitude 07° 21′ W

In addition, as there might well be insufficient time for the bridge to communicate the ship's position to the operator, a slate was kept in the wireless-room on which the position was written up every half hour. The above message would have taken less than half a minute to transmit, so the receiving operator needed to be very much on the alert.

It is impossible to say how many such distress calls were sent out during the conflict, but the number must have been very high indeed. Even under peace conditions the number grew considerably and in 1948 British Post Office stations dealt with no fewer than 281, though probably many of these would have been from small craft.

4
Rival Systems

So far we have considered only the Marconi spark system of wireless telegraphy, the early development of which took place mainly in England and, to a lesser extent, in other European countries. On the other side of the Atlantic development was taking place along quite different lines.

It may be said that the history of American wireless or radio dates from a discovery by Thomas A. Edison in 1888, that if the filament of an electric lamp is heated, electrons will be emitted from it, a phenomenon that became known as the 'Edison effect'. However, Edison did not investigate the significance of this discovery and it was left to Fleming and De Forest to pursue it.

The history of wireless telegraphy and radio telephony in the United States during the first two decades of the twentieth century is largely the story of the careers of three outstanding investigators, though there were others who must not be forgotten. The first of the three was Reginald Aubrey Fessenden, who was born in Canada in 1866. After graduating he joined the Edison organisation, first as a tester and soon as chief chemist at the Edison Laboratory at West Orange, New Jersey. Later he became professor of electrical engineering at Pittsburgh University. In 1910 he was appointed by the US Department of Agriculture to study the practicability of using wireless for the collection of meteorological information, thus bringing wireless telegraphy and meteorology together for the first time.

At that time all wireless was based on the spark system developed by Marconi and others. In this the waves were radiated in short bursts or trains of rapidly diminishing amplitude and repeated at very short intervals. Fessenden conceived the idea of radiating a continuous series of waves all of equal amplitude, the system which became known as continuous wave, or cw. His main interest was not in wireless telegraphy but in radio telephony, which is only possible with cw.

Fessenden's idea was to produce the necessary oscillatory current not by a series of condenser discharges across a spark gap but by charging the aerial directly from a radio frequency alternator. At that time no suitable alternator was in existence. The highest frequency obtainable would have been in the region of 500 cycles per second while what he needed was something of the order of 50,000 cycles (50kHz). Fessenden approached the General Electric Co who assigned one of its staff, a Swedish engineer by the name of Alexanderson, to tackle the problem; Alexanderson succeeded in producing a satisfactory machine which became known as the Alexanderson alternator, and which, when further developed, became the source of power for several very high-power stations.

Fessenden now had a transmitter but the only detector available was the coherer and that was not suitable for telephony. He now set about designing something better and soon he produced the electrolytic detector, a simple and efficient device which was extensively used for many years. However, it too was unsuitable for receiving continuous waves. For this Fessenden devised the beat or heterodyne system. It is impossible to hear radio frequency signals as received by the aerial because their range of frequency is far above the range of human audibility; a satisfactory note for receiving morse signals is of the order of 1,000 cycles. Using a tuning fork, Fessenden generated locally oscillations having a frequency of 1,000 above or below the incoming frequency and mixed the two. The result was that every 1,000 cycles the local and the incoming frequencies coincided, producing an audible note. This was the beat system which is universally employed for receiving CW today, and it is one of the very few of the early basic inventions which have survived and are likely to continue to do so.

Using his new alternator, with the output modulated with a carbon microphone, and his electrolytic detector, Fessenden succeeded in 1902 in transmitting the human voice over the air, the distance accomplished being one mile. This was an historic achievement but in 1906 he achieved a 'first' that was even more portentous. On Christmas Eve, using a 1kW Alexanderson alternator on a wavelength of 7,000m (42kHz), he made the first ever broadcast of a programme of speech and music. It included the voice of a woman singing, a violin solo, and an address. It was heard by many listeners, mainly professional wireless operators, over distances of up to several hundred miles. To them it sounded uncanny, something out of another world. This was just fourteen years before the opening of KDKA and sixteen before the BBC started regular broadcasting.

Fessenden was one of the great inventors in this field; he was extremely prolific, even more so than De Forest, being awarded five hundred patents. The most important of these concerned the radio frequency alternator, the electrolytic detector, and the heterodyne receiver. He launched the National Electric Signalling Co to exploit his inventions but like De Forest he had trouble with his backers and did not profit from his talents as much as he should have done.

Another high frequency alternator was invented by Dr Goldschmidt in Germany. This was an extremely ingenious machine and provided power for many German-built stations.

The most widely used unit for CW and telephony before the introduction of the valve transmitter was the Poulsen arc, invented by Poulsen, a Danish engineer working in the United States. It was more satisfactory for low power than the alternators and had other advantages, but it could never have formed the basis of modern radio broadcasting. (Descriptions of the systems mentioned above will be found in Appendix A.)

Lee De Forest, the second of the three, was born in Council Bluffs, Iowa, on 20 August 1873, so he was one year older than Marconi. His father, a minister, wanted him to follow him into the same profession and sent him to Yale but De Forest was much less interested in theology than in science and his reading at Yale consisted of a strange mixture of holy writ and the *Patent Office Gazette*. He became deeply interested in the experiments being carried out with Hertzian waves and while still at Yale began conducting his own. On leaving he entered the service of the Western Electric Co in their telephone department, continuing his experiments in his spare time, and in 1902, with the backing of a Wall Street financier, Albert M. White, he formed the American De Forest Wireless Telegraphy Co to exploit his inventions.

The first big success of this company was to obtain an order from the US War Department, which had decided it was time to do something about wireless telegraphy, to erect two experimental coast stations at Fort Mansfield and Fort Wentworth. These successfully took part in the US naval manoeuvres of that year, the first time wireless was used in the United States for military purposes. Soon afterwards De Forest was given a contract to erect five powerful stations for the US Navy at Pensacola, Florida; Key West, Florida; San Juan, Puerto Rico; Guantanamo, Cuba and Colon, Panama (Fig 4). He also fitted numerous ship and other stations.

During the period 1902-6 De Forest introduced a number of innovations in the field of spark telegraphy but none was of outstanding importance. However, in 1906 one of his directors, General Dunwoody, invented the carborundum crystal detector which was very widely used for many years and is still far from obsolete.

Meanwhile, De Forest had been greatly interested in the 'Edison effect' and, when it appeared, the Fleming diode. Fleming had encircled the filament of a lamp bulb with a metal shield or plate with an outside connection sealed through the glass. With this arrangement he found that current would flow between filament and plate in one direction but not in the other, so that the device formed a sensitive and stable detector. De Forest started experimenting with this and in 1906 produced his triode, or three electrode valve. In this a mesh or spiral of very fine wire was placed between the filament and the plate of the Fleming valve,

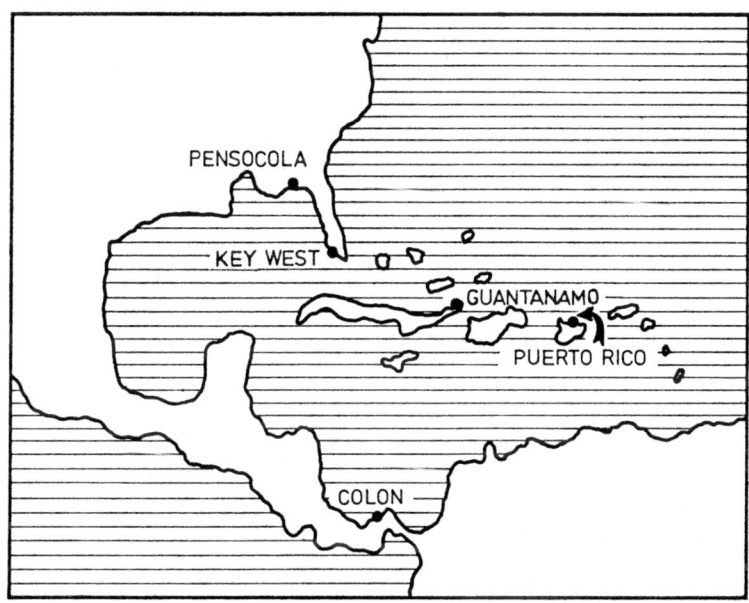

Fig 4 First US naval stations

enabling the current passing between them to be controlled and thus allowing the valve to be used not only as a detector but also as an amplifier of signals. From this point modern electronics may be said to date.

This invention has been compared in importance to the invention of printing. There is no doubt that it triggered off a revolution in electrical engineering, making many things possible which were quite impossible without it, not the least being broadcasting and television. However, today, less than three-quarters of a century later, we can see that whatever its short-term impact, it is not likely to have anything like the life of printing, for the valve and all its vast technology are rapidly being displaced by the semi-conductor with its own technology.

De Forest called his valve an Audion and applied for a patent at the end of 1906. At the same time White, with the connivance of other directors, carried out a Stock Exchange transaction which converted the American De Forest Wireless Telegraphy Co into the United Wireless Co of America. De Forest refused to have any part in this very dubious transaction and he resigned from the company, being paid the ridiculous sum of $1,000 for his patents, though the directors graciously allowed him to keep his Audion patent, then pending, thinking it of no value.

De Forest had been taking some interest in the Poulsen arc system, and he now dropped wireless telegraphy and turned his attention to radio telephony, using the arc for transmitting and his Audion for receiving. In order to develop and market for his system of RT he launched the De Forest Radio Telephone Company, though perhaps 'launch' is not quite the right word, for when he tried to obtain adequate capital for the enterprise he found that it was not only his old directors who thought little of his Audion. Nevertheless, De Forest started well in his new field. In the spring of 1907 he installed RT on the Lackawana Railroad's Hudson River ferryboat, *Bergen*, and at the Hoboken and 23rd St terminals, enabling communication with the ferry to be effected at any time. Then in September of that year he gave a demonstration of RT aboard the *Virginia* and the *Connecticut*. This was the first use of RT on naval vessels and it was so successful that Admiral Soames, who was about to take a squadron on a round-the-world cruise, insisted that all his ships be fitted. The range guaranteed was only eight miles but more was achieved in service.

Early in 1908 De Forest crossed with his wife to France and obtained permission to set up a RT transmitter at the foot of the Eiffel Tower; with an aerial suspended from the tower he gave a demonstration which took the form of a long programme of gramophone records. It was a spectacular success, the broadcast being heard over a wide area, one listener actually picking it up clearly in Marseilles, five hundred miles away. In 1910 he again attracted a great deal of attention by erecting a transmitter on the roof of the Metropolitan Opera House, New York, and on 12 January broadcasting direct from the stage the voice of Caruso. This was heard over a large area, many people sharing headphones to listen to it.

Meanwhile, the development of the Audion was not proceeding as rapidly as one might have expected. A good many of the tubes were sold to amateurs and experimenters but it had no great impact on the field of wireless and it is safe to say that De Forest himself did not realise the full potentiality of his invention. It was even some time before it occurred to him that several Audions could be used in a receiver for amplification, instead of just one. It was in fact still a very inefficient device, compared with modern valves, having an amplification factor of no more than 3. In 1912, however, he found that by using three Audions in cascade he could amplify a signal by $3 \times 3 \times 3$, that is to say, 27. It then occurred to him that it would make an ideal device for amplifying long-distance line telephony and with this in mind he approached the American Telephone & Telegraph Co. They tested it and found that it was just what they were looking for, a means of making coast-to-coast telephony possible. They offered to buy the Audion patent.

De Forest reckoned that he would be able to get half a million dollars for the patent, and according to one account the company itself assessed its value at that figure. Nevertheless, they managed to persuade him to accept a paltry $50,000; once again De Forest had proved no match for big business. However, all was not plain sailing for the AT & T, for it soon found itself involved in a long and costly patent suit with the Marconi Company, who claimed that the Audion patent was an infringement of the Fleming valve patent, which they owned. The outcome was not finally settled until 1916, when the American Supreme Court ruled that when the Audion was used as a detector it infringed the Fleming patent but that it did not do so when used as an amplifier or for generating oscillations.

This created a very difficult situation for designers, a situation which was further complicated by the invention and patenting by Edward H. Armstrong of the regenerative or reaction circuit. In this the output of the detector valve (Audion) was fed back to the grid circuit, and so amplified, greatly increasing the strength of signal received, though when carried too far causing instability which in turn set up the whistles and howls which older listeners will remember as a feature of broadcasting in its early days.

The result of the Supreme Court's ruling was that nobody could construct and market a sensitive, efficient valve receiver without infringing somebody else's patent. It looked very much as though it would be impossible to resolve the difficulties without protracted negotiations and possibly further litigation.

At this point world events took a hand. World War I had already begun and when the United States entered it early in 1917 there was an immediate demand for immense quantities of all kinds of wireless equipment. Now there was no time for wrangling over patent rights; the Secretary of the Navy ordered all manufacturers to use any patents they might require and gave them an indemnity against all claims for infringement.

We have said that the development of the Audion was slow and indeed it is strange that a man with such a creative mind as De Forest should not have realised the full potentiality of his great invention. It is particularly surprising that, even if it had never occurred to him that it was capable of producing oscillations, he did not stumble on the fact accidentally, as for several years after it came into general use a great deal of thought and inventiveness went into preventing it oscillating when not required to do so. As it was, this property was discovered practically simultaneously by three other investigators, Armstrong in America, Meisner in Germany and Franklin in England.

Rival Systems

The third of our three giants, Edward Howard Armstrong, was born in New York on 18 December 1890, so that he came on the scene somewhat later than did the others. However, he wasted no time in getting started and before he left Columbia University in 1913 he had already embarked on his spectacular career as an electronics engineer, having in 1912 invented the feedback circuit, which was a feature of most radio receivers during the next ten years or so and which incorporated the principle of self-oscillation which enables the valve to be used for transmitting CW.

When the United States entered World War I Armstrong joined the US Signal Corps, and in 1918, while serving as a captain, he invented the super-heterodyne circuit. This he followed in 1920 with the super-regenerative circuit. His fourth—and in the opinion of some people his greatest—invention was frequency modulation (FM). To demonstrate this he set up the first FM station at Alpine, New Jersey, in 1937. FM is now used in all VHF broadcasting and TV. Like so many other inventors in the radio field, Armstrong was plagued with patent litigation and although at one time he acquired a considerable fortune he lost most of it and—particularly tragically for one who had contributed so much to the stock of human knowledge—committed suicide in 1954.

De Forest has been called the Father of Radio but one cannot help wondering whether this title is justified. Fessenden was the first to produce both a practical transmitter and receiver of CW as well as the first to transmit an actual programme of words and music. De Forest turned the diode into the triode, making it possible to amplify received signals, but he did not develop it as a transmitter. As for Armstrong, radio broadcasting would have been possible without his inventions, but only just. The truth is that no one man can fairly be called the Father of Radio, any more than there was ever just one inventor of wireless telegraphy. Many men combined to make the whole, each building to a large extent on the work of others.

As a result of the pioneer work done first by Fessenden and

Poulsen and then by De Forest, interest in radio spread rapidly throughout the United States. From 1910 onwards several big corporations, prominent among them General Electric, AT & T and the Bell Systems Laboratory were very active in this field. However, they all observed a great deal of secrecy and it is impossible to say anything here about the lines along which they were working.

In addition to the professional engineers and the big corporations a vast number of amateurs, young and old, had become interested in radio and were engaged in more or less serious experimentation. Some confined themselves to receiving, being mainly interested in station identification, while the more adventurous set up transmitters. The activities of these latter often amounted to unofficial broadcasting, their programme material consisting of records and the vocal and instrumental efforts of their friends. With them there was very little distinction between straight radio telephony and broadcasting.

At that time there were no production lines with complete domestic radio receivers as their end product and anyone who wanted a receiver had to make it himself. Materials were easy to obtain; such items as an Audion were comparatively expensive but a simple crystal receiver cost very little to make. The only difficult component to obtain was the telephone but many experimenters solved that with a visit to a public call box.

The vast multiplicity of listeners presented no problems but with the amateur transmitters it was a different matter. There were no regulations governing the use of the ether; anyone could set up a transmitter and transmit for as long as he liked on any wavelength that took his fancy, using as much power as he could afford. Many amateurs took the view that the air was free for all and acted accordingly. The result was chaos. Often the transmitters were very flatly tuned and caused a great deal of interference, not only to each other but to naval and commercial stations, the Brooklyn Navy Yard being particularly troubled by them. Eventually, in 1912, the US Congress passed the Radio Act.

This laid down that all transmitters and receivers must be licensed by the Secretary for Commerce, but it went on to enact that the secretary could not refuse a licence to any American subject. As a result of this very strange provision, the Act did practically nothing to reduce congestion.

When amateur activities were shut down on the United States becoming a belligerent in 1917 it was estimated that throughout the country there were 8,500 amateurs licensed to transmit. Considering how unselective were the receivers then in use one can see what a nuisance they must have been to important services, and that in coastal areas there would have been a serious risk of them interfering with distress calls. No doubt some of the many experimenters made useful contributions to radio technology, but their main importance lay in the fact that they formed the nucleus of an audience for broadcasting when it came to be developed commercially, and that the more knowledgeable among them became very useful as personnel for the radio sections of the armed services when war broke out.

5
Trans-Ocean

The impact of wireless telegraphy in the maritime field was so enormous because it offered a service which was of great value and which could be achieved in no other way. When considered as a means of communication between fixed points, it was low cost that was its attraction. It was soon seen to be ideal for extending a telegraph service to small islands in situations where the traffic would be too light to warrant the cost of an expensive submarine cable; quite early in the 1900s stations were erected to link up some of the Scottish islands and in March 1901 a public service was opened between the islands of Oahu, Kousi, Molaki, Maui and Hawaii in the Hawaiian group. As technical improvements were made and range extended, pairs and groups of such stations sprang up all over the world, particularly in the less-developed areas where difficult terrain such as forests and mountain chains combined with low population density made the construction of land lines too costly.

The position with regard to long-distance point-to-point services was quite different. By the beginning of the twentieth century there was already an extensive international cable network providing rapid and reliable communication between important points. At first sight, therefore, there did not seem to be much need for wireless. Marconi, however, did not share this view. At a very early stage in his work he envisaged great possibilities for the new system of communication and even before he had achieved a range of a hundred miles he was dreaming of spanning the Atlantic. Nor did he stop at dreaming. As soon as he had established the practicability of his system over useful distances, he set about achieving his goal, even while his only detector was the far from satisfactory coherer.

When his intention became known the 'experts', as we have seen, declared that what he proposed to do was impossible, maintaining that the Hertzian waves would shoot off tangentially from the earth's surface and pass high above the receiving station on the other side of the ocean. We now know that waves of the length Marconi was using are reflected back to the surface by the ionised upper atmosphere. At the time, though, it was clearly necessary to determine experimentally exactly how they would behave.

Marconi's first step was to erect the famous high-power station, Poldhu MPD in Cornwall. This was equipped with a transmitter deriving power from a 25hp engine driving a 2,000V alternator, the output of which was transformed up to 20,000V. Delay was caused when a gale blew down the somewhat insubstantial masts, but these were replaced by four lattice towers and towards the end of 1901 everything was ready for the experiment on which the whole future of long-distance wireless telegraphy depended.

Marconi, with two assistants, travelled to St John's, Newfoundland, where he at once established a temporary receiving station with an aerial supported by a kite; he had arranged that Poldhu should transmit a series of the letter s continuously according to an agreed time schedule. On 12 December 1901 both he and an assistant distinctly heard the repeated s though as the detector was a coherer these took the form of clicks in a telephone receiver. The same result was achieved when the experiment was repeated on the following day. Marconi's objective had been achieved and he had shown beyond dispute that the electromagnetic waves followed the curvature of the earth. (This is not, in fact, true of the ultra-high-frequency waves used in space communication.)

Perhaps it is too much to say beyond dispute. The theorists did not take their defeat lying down, and declared that the letter s in morse could easily be confused with static and that this was all Marconi had heard. Indeed Marconi was, as we know, a telegraphist, and it does seem surprising that he should have chosen to use this one letter for such an important experiment; one would have expected him to use a more distinctive letter, such as v or c, or perhaps a short word. And unfortunately only he and an assistant heard the signals.

One often reads that Marconi sent the first wireless message across the Atlantic in 1901. He did not. What was sent was only a signal, which was all that was necessary for his experiment. Possibly he could have received a whole message from Poldhu, but he did not. Marconi confirmed his results shortly afterwards by receiving signals from Poldhu aboard the *Philadelphia* at a distance of 2,100 miles. These were recorded on paper and their reception was attested by the captain and other witnesses. However, even this did not finally silence all those who doubted his claim for the St John's experiment—far from it. In fact even today there are sceptics.

Marconi had intended that if he was successful he would establish a permanent transatlantic station in Newfoundland, and he saw what looked like a suitable site soon after his arrival. However, he had no sooner announced the success of his experiment than the Anglo-American Cable Co, quick to realise the danger to its business that could develop from a system that could achieve transatlantic telegraphy without having to lay and maintain an expensive cable, claimed that it owned monopoly landing rights in Newfoundland and threatened legal action unless he abandoned his plans immediately. Events have shown that in taking this action it showed greater perspicacity than had one of its rivals, the Commercial Cable Co, when it had earlier dismissed with contempt the possibility of competition from wireless.

Marconi did not waste time contesting the Anglo-American Co's claim but promptly crossed into Canada in search of a site there. He was warmly welcomed by Canadian government officials and with their help quickly found at Glace Bay on the coast of Nova Scotia an ideal site for his station. He was promised financial support by the Canadian government and he lost no time in erecting an experimental station; he also erected a transmitter at Cape Cod. Experiments were carried out between these two stations and Poldhu, and on 15 December 1902 Marconi sent the first transatlantic wireless telegram to the London *Times*:

HAVE HONOUR TO SEND THROUGH 'TIMES' THE FIRST TRANS-ATLANTIC MESSAGE TO ENGLAND AND ITALY.

On 3 January 1903 messages were exchanged between King Edward VII and President Theodore Roosevelt.

After further experimental work it was decided to erect a pair of stations for a permanent commercial service at Glace Bay and Clifden, Ireland. The erection of the Clifden station began in 1905, and the circuit and equipment differed substantially from anything previously seen. The usual system, in which current was drawn from an alternator and transformed up to the voltage needed to charge the condenser which provided the oscillatory current for the aerial, was abandoned; instead a 16,000 V accumulator battery provided current which was broken up by a very high-speed interrupter and which charged the condenser without any transformer. This gave a purer note. The accumulator batteries were the largest ever constructed.

These stations incorporated Marconi's directional aerial patented in 1904, which marked a substantial advance in long-distance wireless (Fig 5). It comprised an elevated horizontal wire section with a down lead at one end; the inverted L thus formed was erected with the horizontal direction parallel with the line joining the home and distant station and with the free end pointing away from the distant station. Such an aerial produced the best results for both transmitting and reception along the line of the two stations.

On 27 October 1907 the two stations were opened for a limited public service but then in August 1908 the Glace Bay station was

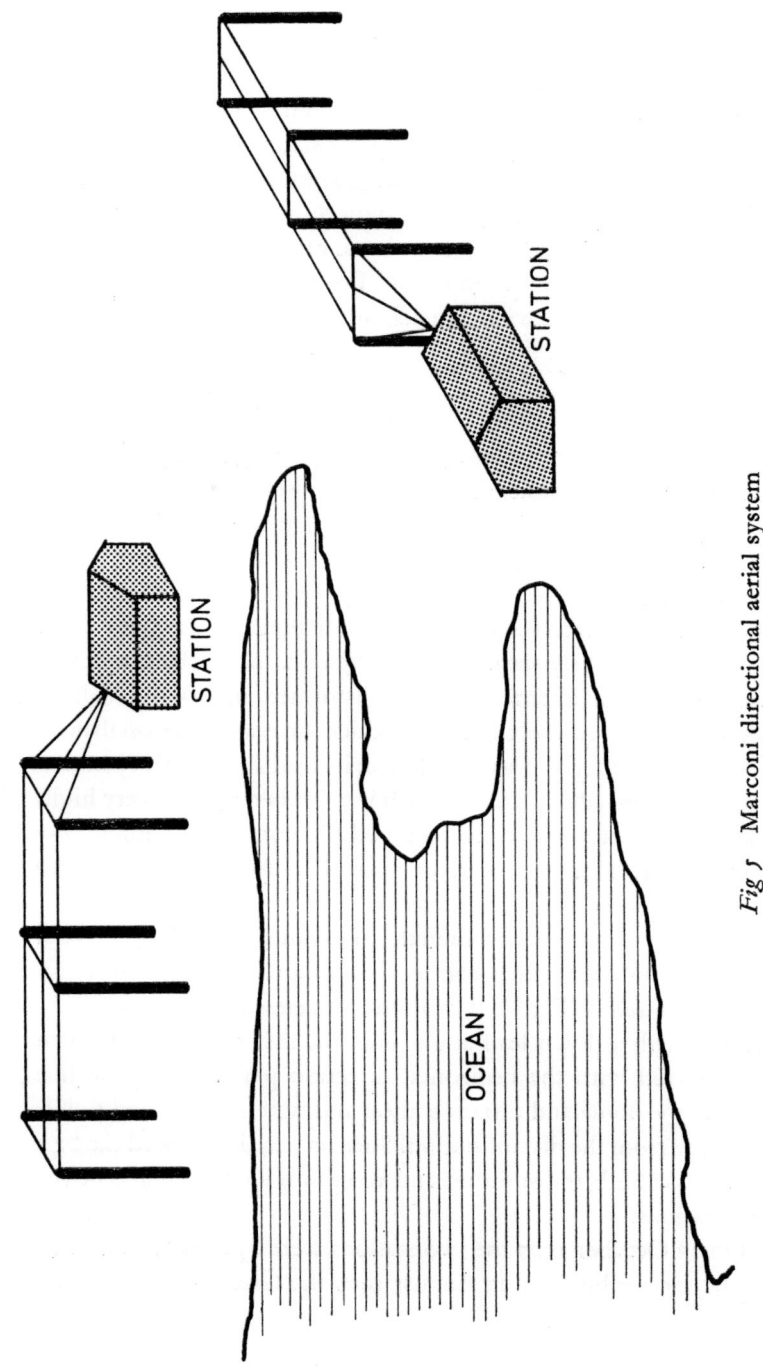

Fig 5 Marconi directional aerial system

destroyed by fire. It was promptly rebuilt and on 23 April 1909 it was back in service, this time fully open to the public. On this date the battle between long-distance wireless and the cables may be said to have been joined, and with the Marconi Company substantially cutting the cable tariffs it began to look as though the Anglo-American Co's fears were justified.

Low cost was not the only advantage which wireless had to offer for international communication. For many countries cables and international land lines had a very serious drawback: part or all of any route might be under foreign control. Even in peacetime this could have disadvantages, which were particularly serious for a landlocked country with no control over the tariffs applied to its communications with the outside world, who could also probably do little or nothing about delays due to inefficiency; there might even be interception of messages. Even countries with coastlines might have no control over the cables, which could be owned by foreign (very often British) companies. In wartime these difficulties might be greatly increased by enemy action. Cables could be cut, as happened extensively in the Mediterranean during World War II. And hostile or merely unfriendly countries could cut off communication completely.

An extreme example of the consequences of foreign (in this case, British) ownership of cables is recorded in the diary of Sir Maurice Hankey, for many years secretary to the cabinet, and quoted by Stephen Roskill in his *Hankey*. During the early part of World War I the German Foreign Office routed its cables to Washington through its Swedish Embassy and from Sweden they were actually routed via England. On the way they were diverted through room 40 at the Admiralty, where their code was cracked. As a result the British Government was kept well informed regarding the Wilhelmstrasse's exchanges with Washington.

Wireless was quite free from these troubles. Its signals could pass unhindered over hostile territory direct to the country of destination; furthermore, its installations could be more easily defended. A wireless station sited away from the coast was much less vulnerable than a cable which could only be defended by complete mastery of the sea.

With the Marconi transatlantic service in full commercial operation these advantages became clear to everyone. Marconi was of course fully alive to them. Even before Clifden and Glace Bay had been opened for business he was looking for fresh fields to conquer. At that time Great Britain was at the height of her imperial power, with colonies and dependencies scattered all over the world, and in 1906 the Marconi Company suggested to the Colonial Office that it should be given a contract to construct a chain of high-power stations to link these with the motherland. The Colonial Office did not agree to the suggestion and was still unsympathetic when the company repeated the proposal in 1909. However, the international situation was changing. War clouds were beginning to gather and the feeling was growing in high places that there was an urgent need for improved intercommunication between the various units of the Empire. A large part of the international cable network was controlled by Britain but in wartime cables could be cut, as everybody knew, and it would be difficult to protect the whole length of one, let alone several. In 1911 the construction of an Imperial Wireless Chain was approved, the contract being awarded to the Marconi Company.

Before work on the proposed stations could be started there blew up a political row which became known as the Marconi scandal—quite unjustifiably, as it involved no member of the Marconi family nor any action of the Marconi Company. It is important to give some account of it here in order to correct any wrong impression which the present-day reader may have formed from casual reading, and also because it delayed work on the chain so that it was not ready for use when World War I broke out. Interest in the affair centred not round the company nor around technology, but around the highly placed public figures involved. These were all members of the Asquith Liberal government then in power, so it is not surprising that the Tory press made the most of it. It was a complicated business, and the following brief summary is based mainly on the account given in Randolph Churchill's *Winston S. Churchill.*

The Marconi Company was, as we have seen, granted the contract to construct the stations and, consequent on the signing of it, the ordinary shares of Marconi's Wireless Telegraph Co Ltd rose spectacularly from £2 8s 3d (£2.41) in August 1910 to £9 on 2 April 1912 (that is, before the boost given to them by the *Titanic* disaster.) There then followed newspaper attacks initiated by a paper called the *Eye-Witness* in which it was alleged that Sir Rufus Isaacs, the Solicitor-General and brother of Godfrey Isaacs, managing director of Marconi's Wireless Telegraph Co, David Lloyd George, Chancellor of the Exchequer, and Herbert Samuel, the Postmaster General, had conspired before the signing of the contract to purchase shares in Marconi's Wireless Telegraph Co and profit by their expected rise on the Stock Exchange. Things soon reached the stage when questions were being asked in the House of Commons and a select committee was set up to investigate the matter.

When it had been first raised in the Commons both Lloyd George and Rufus Isaacs denied that they held any shares in the company, but later they revealed that they had holdings in Marconi's Wireless Telegraph Co of America, an entirely independent concern. Meanwhile, as the result of a rumour that he was involved, published in the *Financial News*, Winston Churchill was summoned to appear before the committee. He was extremely indignant and issued a categorical denial that he held or ever had held any Marconi shares.

The select committee cleared everyone concerned. It reported:

> So far as we are able to ascertain, no cabinet official or Member of Parliament has been influenced by any interest he may have had in any Marconi or other undertaking connected with wireless telegraphy, or has used any information coming from official sources for the purpose of speculation.

Marconi himself was greatly upset by the affair. It must have been extremely unpleasant for him as he went about London constantly seeing the newspaper placards bearing the words MARCONI SCANDAL and appearing to impute dishonourable conduct to himself and his business associates, when in fact they referred to something with which he had nothing whatever to do.*

It is pleasant to be able to record that because of the attitude towards the affair adopted by the Prime Minister, Herbert Asquith, the careers of none of those involved suffered, and they all went on to give very valuable service to their country. However, consequent on the affair, the government decided to reconsider the contract for the imperial chain, and it set up a technical commission under Lord Parker to consider which was the best system for it. It seems strange that this had not already been done.

By this time there were several contenders for the contract. There were two radio frequency alternators capable of delivering alternating current at a radio frequency direct to the aerial: the German Goldschmidt machine and one developed by E. F. W. Alexanderson in America. There was also the Poulsen Arc system, which had produced some promising results. Against these the Marconi Company offered its timed-spark system.

The commission published its report in April 1913. It was quite definite. Referring to the timed-spark system it said:

> The Marconi system is the only system of which it can be said at present with any certainty that it is capable of fulfilling the requirements of the imperial chain.
> ...The only continuous high-frequency generator (cw) which we have yet seen tried with any success over long distances is the Marconi high-frequency machine.

The Marconi Company was given the go-ahead, but the delay resulting from the squalid intrigues of the politicians and the newspaper sensationalists had serious consequences. Very soon the outbreak of World War I forced suspension of the work, and the imperial chain was not ready when it was most needed. By the

* It was later stated that neither Lloyd George nor Samuel made anything out of their dealings in American Marconi shares and actually lost a small sum.

time hostilities had ceased and it was possible to resume work, technology had made immense strides and the plans were out-of-date. However, as we shall see later, the need for an imperial chain persisted, and fresh plans were prepared.

Germany had also seen the need for a worldwide chain of stations. While the start on the British chain was being held up, she set about the construction of an equally ambitious chain based on the giant station at Nauen near Berlin, then the most powerful station in the world; with powerful stations in all her dependencies she pressed ahead with the work, and when war broke out the chain was complete and in service. What use she was able to make of it we shall see later.

The year 1913 was the last full year of many things. For us it was the end of the pre-valve era.

As we have seen, Dr J. A. Fleming had introduced his two-electrode valve in 1904, and although it was used in a number of installations it cannot be said to have had a very great impact on wireless telegraphy. Dr Lee De Forest had transformed it into the three-electrode valve, but although it was being fairly widely used in the United States it had in the seven years up to 1913 made hardly any impact at all in Britain. Looking back, it seems incredible that it took so long for its potentialities to be realised. However, the war very quickly changed all that.

1913 seems a good year in which to take stock. How far had wireless in fact progressed? Technically it had travelled a long way. Nevertheless, if we exclude a few high-power stations it still relied for transmission on the spark system and for reception on the magnetic detector, the crystal detector or the electrolytic detector, and very little on valves. A study of *The Yearbook of Wireless Telegraphy and Telephony* for 1913 gives us a clear picture of the state of wireless telegraphy at this point. The sheer size of the book—564 pages of editorial matter plus 54 pages of advertisements gives some idea of the progress that had been made and this is emphasised by the list of stations then in operation.

TABLE 3

Land stations operating in 1913

	Great Britain	United States	World
Commercial	34	44	
Naval	18	8	
Total	52	52	approx 1,650

'Commercial' includes coast stations, stations engaged in point-to-point working, and lightships. In the case of most countries, including the United States, it is not possible to separate these. There were ten coast stations round the coasts of the British Isles operated by the Post Office and communicating with merchant shipping. These included Cullercoats, North Foreland, Niton and Malin Road, all of which are still in the same position, together with Bolt Head, Caister-on-Sea, Crookhaven, Seaforth, Lizard, and Rosslare, which have either been moved or closed. There were no British Army stations.

There are some notable absentees from the list of countries possessing stations, among them China, Turkey and numerous British crown colonies. In contrast to the rapid development of radio in Great Britain and the self-governing dominions, the dependencies controlled by the Colonial Office lagged far behind. Of these the most remarkable omissions are the great ports of Singapore and Hong Kong. And neither Gibraltar nor Malta, both situated right on the main route to the east, had a commercial station. One cannot help thinking that there was a connection between this fact and the failure to provide time signals in British overseas territories.

High-power stations for regular trans-ocean services are shown in the 1913 Yearbook as being under construction in several countries: the United States, five; Britain, Norway and Argentina, one each; and the map shows the position of several such stations projected for the imperial wireless chain. However, it appears that

Page 85 (*right*) A trench WIT station on Gallipoli in World War I; (*bottom*) a Marconi ½RW pack set at Falahiyeh, Mesopotamia, in World War I. The rotary spark, with the cover removed, can be seen to the left of the picture

Page 86 (*top*) An American soldier operating a trench set on the Western Front; (*bottom*) a ship's radio-room, believed to be that of the *Olympic* in 1917. The equipment visible includes the induction coil, the magnetic detector, the multiple tuner and the value receiver

the only long-distance service actually in operation was that between Clifden, Ireland, and Glace Bay, Canada, though there were a good many regular services in operation over shorter distances.

The list of land stations is followed by another list giving some three thousand ship stations but this figure must be taken as approximate, as it was increasing very rapidly and in addition one suspects that a good many warships were not included.

There are notable omissions from the Yearbook. The three-electrode valve is hardly mentioned. Radio telephony, too, although it is included in the book's title, receives little attention, and an article on it dismisses it as something not yet practicable. This despite the fact that considerable advances with it had been made in the United States, though not in Great Britain. We can only ascribe these two omissions to British insularity. One omission more noticeable than any other is that of the use of wireless for purposes of entertainment, as experimental broadcasts of music and speech were already taking place in the United States, though not, again, in Great Britain.

The Yearbook shows that by 1913 wireless had proved a boon to shipping and was beginning to cut in on the preserves of the cable companies, but as yet it had not made any impact on the lives of the majority of people.

6
Preliminaries to War

The first military use of any form of electrical signalling was in the Crimean War. When that conflict broke out, the European telegraph network was still far from complete and the British and French forces landing on Russian soil found that the nearest point from which telegraphic communication was possible with their respective capitals was Bucharest. Fortunately, that city was then under Turkish, not yet Russian control, and Turkey was an ally. Accordingly a land line was constructed from Bucharest to the Black Sea port of Varna and from there a submarine cable was laid to Balaclava in the Crimea. On 31 May 1855 telegraphic communication was established between London and Paris on the one hand and British and French GHQs on the other, and for the first time armies fighting in a distant field were under the immediate control of their home governments. The Russians for their part constructed a special line linking St Petersburg (Leningrad) with Sevastopol.

The service, at any rate on the allied side, was not an outstanding success. Whitehall pestered the British GOC with an endless series of telegrams concerning trivial matters, whilst the French general received unwelcome directions from the Emperor Napoleon III on how to conduct the campaign. Nor was it easy to prevent frequent interruptions of the service. The unprotected insulation of the somewhat primitive submarine cable was liable to damage, and on land the troops found pieces of telegraph wire handy as pipe-cleaners so that any unburied section of line was likely to have a short life.

Not long after the close of the Crimean War the American Civil War broke out. During that unhappy struggle, which was fought over a wide area, the use of telegraphy for military purposes was developed considerably and much useful knowledge was gained.

Line telegraphy was used by the British Army in several small wars during the second half of the nineteenth century, and by the outbreak of the South African War it was fully established as a means of signalling.

The first use of telephony for military purposes was in the Egyptian War of 1882, at which date it had only been invented some six years. It was therefore not much understood in military circles and for several years it was only handled by the signallers. Apparently the officers of that day considered it beneath their dignity to use it.

By 1898 Marconi had developed his wireless telegraph to the point when it was possible to give a demonstration to parties likely to be interested. He did this on Salisbury Plain before representatives of the Post Office and the two armed services, the Royal Engineers sending an officer to represent them. The Engineers were much impressed by his report and promptly established an experimental wireless telegraphy section at Aldershot. Major-General R. F. Nalder tells us in his *History of the Royal Corps of Signals* that a wireless set was developed which was sent to South Africa for use in the field there. Unfortunately it was not found to be satisfactory. At the same time the War Office recognised the Marconi system of wireless telegraphy for use in the Army. Soon after the outbreak of the South African War it purchased six sets from the Marconi Company and these were despatched to the Cape along with six Marconi engineers. However they were not a success. The War Office sent out masts with the sets but these would not stand up to the high winds experienced on the veldt. Nor was it found possible to obtain suitable poles locally. The sets were eventually transferred to the

Navy and installed on some of the ships blockading the coast of South Africa where they gave good service.

There was a third attempt, independent of the two already mentioned, to adapt the new invention to military use during the South African War. The beleaguered garrison of Ladysmith, with no means of communication with the outside world except an uncertain heliograph (an instrument which signalled by flashing light reflected from the sun), attempted to improvise a wireless station. The officer responsible must have been unusually technically minded and well-informed about current developments but it can hardly have been possible to have found the necessary material in such a place at that time and the attempt failed.

Apparently the Boers also experimented with wireless telegraphy. According to Erik Barnouw in his *History of Broadcasting in the United States*, on 3 December 1899 the *New York Herald* reported that British forces in South Africa had captured from the Boers some wireless equipment which had been made in Germany and closely resembled Marconi apparatus.

The Russians were the next to use wireless telegraphy on the battlefield. This was in 1904-5 during their war with Japan in the Far East. Their handling of it was in keeping with the general incompetence which characterised their conduct of the campaign, for they eschewed the use of code, thus communicating much vital information concerning their movements to the enemy. That the Japanese were able to intercept it shows that they too had military wireless telegraphy at a very early date. Knowledge of this—quite unnecessary—leakage of secret information prejudiced military thinking against wireless telegraphy in both Britain and Germany right into World War I.

We now come to the almost forgotten Balkan War of 1912, in which Greece, Bulgaria, Montenegro, Serbia and Rumania united to throw off the yoke of Turkey, and which resulted in the western frontier of the once mighty Ottoman Empire being pushed back almost to the Bosphorus. This conflict has a special interest for us because it was the first in which wireless telegraphy can fairly be said to have played a notable part. The Rumanian Army possessed fourteen Marconi 1½kW portable wireless telegraphy stations, and so as we shall see was rather better equipped in this respect than was the more prestigious British Army at that time; and it made good use of them. However, theirs was not the only army possessing the new equipment; the Turks also had at least one station and they not only broke new ground with it but they did so in a most spectacular manner.

The allied armies were quickly successful and soon had the Turkish Army falling back on Constantinople (now Istanbul). In the process a Turkish force under Sukri Pasha was cut off in Adrianople and completely surrounded by the Serbian and Bulgar armies. However, unlike commanders who had previously found themselves in this unhappy situation, Sukri Pasha had no difficulty in maintaining communication with the outside world. By a fortunate chance there was in the fort just one Marconi 1½kW portable wireless station. With this he was able to maintain contact with Constantinople during the whole period of the siege, a very large number of words being transmitted in each direction.

This was the first occasion on which wireless telegraphy was successfully used by a beleaguered force to maintain communication with its own army.

It was not long before the exercise was to be repeated on a vastly larger scale. When World War I broke out in 1914 Germany soon found herself hemmed in on all sides by her enemies. She was never completely surrounded, for throughout the conflict there were neutral countries which formed breaks in the ring, but even so, the allies controlled the international cable network and Germany found wireless telegraphy her only means of communicating with her distant friends without interception or censorship.

We must now return to the development of wireless telegraphy by the British Army. With the rapid advances being made in the technology and use of the new invention in other fields, the army had to consider its position with regard to it; however 'mechanisation' and 'electrification' were words not much used in military circles at that time and outside the Royal Engineers enthusiasm for it was limited. In the early 1900s the British Army had just emerged from an open war of movement in South Africa and senior officers no doubt visualised future campaigns being fought in the same manner. When, therefore, in 1903 it became clear that the introduction of wireless into the army must be considered, it was decided that its only use would be to enable cavalry units operating at a distance to maintain communication with headquarters. Lieutenant-Colonel (later Brigadier) R. Chevenix-Trench in an article published in 1929 in the *Cavalry Journal*, vol XIX, describes the establishment of British Army wireless as an adjunct to the cavalry.

The task of developing suitable equipment was given to the Royal Engineers, who studied the problem at the School of Military Engineering at Chatham. In the following year a Wireless Section of the Royal Engineers was formed at Aldershot in close touch with the cavalry and by 1907 there were two wireless telegraphy companies attached to the 1st Cavalry Brigade. They were regarded with extreme disfavour by senior officers who feared, with complete justification, that if this newfangled means of communication was allowed to develop it would soon upset all their familiar ideas about the conduct of war. However, develop it did, though in the army with extreme slowness. In 1911 the two companies were reorganised as a separate unit named the 1st Wireless Company. This was equipped with a set mounted on two limbers, one of which carried an engine and generator, with a 50ft telescopic steel mast. The whole outfit weighed more than two tons and required six horses to draw it. It took twenty minutes to erect and start working, and it had a range of 50 miles. Compared with the sets available for marine use at the time it does not seem to have been very efficient.

Meanwhile the Marconi Company had developed a pack set for army use, and an officer of the Westmorland and Cumberland

Yeomanry equipped his unit with two at his own expense. It had a range of 15 miles and was so successful that the War Office approved it for army use. However, their approval did not go so far as the placing of a substantial order for the new equipment.

By 1912 the British Army possessed one squadron equipped with three army wagon sets and three pack sets, and a second squadron with three wagon sets and a motor set. This position remained static until the outbreak of war in 1914, so that the British Army took the field possessing only ten wireless stations. This deplorable state of affairs cannot be ascribed to the absence of suitable British equipment. Reference to Table 4 will show that the Marconi Company had developed a whole range of sets for military purposes. No doubt they sold them to foreign armies, but precious few were purchased by the British War Office at a time when a conflict with Germany was looming large on the international horizon.

It is difficult to see how, with such a pitifully small establishment of men and equipment, there could have been any effective practice in the handling of instruments and traffic; nor could there have been any real attempt to foresee difficulties that might be encountered in the field in order to devise ways of overcoming them; and certainly there could have been no adequate study of the effect which wireless would have on both strategy and tactics. If more—much more—had been done, wireless would have made a much better showing on the Western Front during the first few months of the war than it actually did. The blame for this failure must be placed on the higher command for their lack of imagination and hostility to change. Although the fear that wireless telegraphy was vulnerable to interception was a reasonable one, means of overcoming this difficulty were not properly studied.

When we turn to the Royal Navy we find a very different state of affairs. Like the army, the navy sent officers to observe Marconi's demonstrations on Salisbury Plain, and they liked what they saw very much indeed. The navy was at that time much more technically minded than the army, and it never had any doubts about the advantages of wireless telegraphy. Several ships were fitted as early as 1899, and in the naval manoeuvres held in July of that year three warships, the flagship *Alexandria*, and two cruisers, *Juon* and *Europa*, were equipped, exchanging messages up to seventy-four miles. On 24 July 1900 the Marconi Company signed a contract with the Admiralty for the supply of two coast stations and twenty-six ship stations.

The Royal Navy was fortunate in being represented in Whitehall during the period 1903-9 by that redoubtable mariner, First Sea Lord Admiral Sir John (later Lord) Fisher. He was tireless in the work of modernising the navy and, in particular, equipping it with wireless and submarines. Richard Hough in his *First Sea Lord* quotes from his speech at the Royal Academy dinner in 1903:

> Looking at the submarine boat and wireless telegraphy, when these two are perfected we do not know what revolution may come about. In their inception they were the weapons of the weak. Now they loom large as the weapons of the strong. Will any fleet be able to live in the narrow waters? Is there the slightest risk of invasion from them?

In 1903 the Marconi Company entered into an agreement giving the British Admiralty a licence to use all Marconi patents. By 1905 every ship of the British fleet had been fitted, as was the American fleet. A large number of German warships were also equipped. Parallel with the fitting of ships all three navies quickly built shore stations round their home coasts and at bases abroad.

Lord Fisher did not mention the third great new weapon which was to have such an impact on the conduct of World War I. It had in fact not yet appeared on the scene, for Orville Wright did not take off at Kitty Hawk, North Carolina, until several months after that Academy dinner. The wireless set and the aeroplane grew up together, with wireless the elder brother, a relationship which was to continue; for although radio communication today is not dependent on the aeroplane, modern aviation, and particularly space-flying, could not exist without radio control.

The story of wireless in the air goes back to the time before the formation of the Royal Flying Corps and the Royal Naval Air Service, to the time when flying was not a separate service. Flying was not represented at those demonstrations of Marconi's on Salisbury Plain. Nevertheless, the importance of wireless to flying was recognised at a very early stage. The airman's ability to see what the enemy was doing on the other side of the hill and to spot where the artillery's shells were falling completely revolutionised land warfare but this power would have been of little use had he not possessed the means of signalling what he saw back to the ground, without having to wait to land.

It was four years after Kitty Hawk that a wireless set rose into the air for the first time. By that time wireless stations were springing-up everywhere and the shores of the Atlantic were linked by a commercial wireless service. According to Sir Walter Raleigh's *Official History of the War in the Air*, the two technologies first came together in 1907. At that date the British Army had no aeroplanes but it did have some balloons, and it seemed to Captain Llewellyn Evans, commanding the 1st Wireless Company of the Royal Engineers, and Lieutenant-Colonel Cooper of the Balloon School, Farnborough, that there was scope for joint experiments. Balloons had already been used to support aerials and it seemed likely that they could usefully support a good deal more equipment.

The first step was to install a receiver in a captive balloon; Lieutenant C. J. Ashton ascended in it and successfully received signals from the ground. In so doing he became the first man in England, probably in the world, to receive wireless signals in the air. Then in May 1908 an experiment was made with the free balloon *Pegasus*. This too was successful, signals being received from the Aldershot station at a distance of twenty miles. A further step was taken later that month when the first transmission from a free balloon was carried out, with satisfactory results.

The next stage was to try out wireless in an airship. In 1909 Captain P. T. Leroy RE was made responsible for all military wireless experimental work, and when the airship *Beta* was ready for service he had her equipped with a transmitter and a receiver. He went up in her on 27 January 1911 and was able to transmit a number of messages to a ground station at distances of up to thirty miles. Messages were also received but for this it was necessary to stop the airship's engines.

In the following year two more airships became available and it became possible to demonstrate the value of military wireless in a very practical manner. For the Salisbury Plain manoeuvres of 1912 the troops taking part were divided into two forces, the attackers and the defenders, the latter under General Grierson. *Delta* was attached to the attackers, *Gamma* to the defenders, both airships having wireless. At an early stage the *Delta* broke down, leaving only the defenders with an airship; the result is best described in General Grierson's own words, quoted in the *Official History*.

> The impression left on my mind was that they had revolutionised the art of war. As long as aircraft are hovering overhead of troops all movements are liable to be seen and reported, and the first act in war will be to get rid of the hostile aircraft. The airship, so long as she remained effective was of more use to me for strategical reconnaissance than aeroplanes, as being fitted with wireless telegraphy I received her messages in a continuous stream and immediately after the observations had been made. It is a pity that the airship cannot *receive* reports by wireless, but doubtless modern science will soon remedy that defect.

Every day the *Gamma* kept the general fully informed on the enemy's movements. Attempts were made by the latter to screen halted troops but as General Grierson said, 'For troops on the march there is only one certain cover, the shades of night'.

This convincingly demonstrated not only the value of aerial reconnaissance but also the need for instant reporting by wireless. However, it was very soon realised that airships flying low enough for reconnaissance were too vulnerable from the ground and

Preliminaries to War

before war broke out all the army airships were transferred to the the Royal Navy. Thereafter the British Army used aeroplanes for reconnaissance.

Meanwhile the development of the heavier-than-air aeroplane was proceeding rapidly on both sides of the Atlantic. Two-way wireless communication between the ground and an aeroplane in flight was first achieved in 1910 at Sheephead Bay, New York, by McCurdy in a Curtiss aeroplane. At about the same time Robert Loraine, a well-known London actor as well as a keen civilian airman, had his own Bristol aeroplane fitted with a transmitter and at the time of the 1910 army manoeuvres took it up at Larkhill. With it he succeeded in transmitting to a ground station over a distance of about a quarter of a mile. When one remembers the extremely exposed position of the pilot in such early aeroplanes, one sees that it must have been a considerable physical feat to have coped with both the wireless telegraphy and the aeroplane at the same time.

As we have seen, the potentialities of the aeroplane in war were recognised at a very early stage in its development. The first actually to use it in war were the Italians, in their war with Turkey in Tripolitania in 1911, though they do not appear to have used wireless telegraphy with it. In England serious experiments with the use of wireless together with heavier-than-air machines began at Brooklands in the same year, when Lieutenant Donnington-Bungay achieved two-way communication in a Flanders monoplane.

Pilots, still without enclosed cockpits, were very exposed, and this was a source of many of the difficulties which had to be overcome. Among these were the noise from the engine, the slipstream from the airscrew, the difficulty of protecting the apparatus from oil splashes, and above all the danger from fire. In those open-cockpit machines aero spirit was never very far away and the primitive transmitters necessarily produced sparks with their induction coils; there was also always the risk of undesired sparks occurring between some part of the aerial and the airframe. All these were apart from the ever-present difficulties of bulk and weight.*

Meanwhile the Royal Navy was taking an interest in flying. In August 1912 Lieutenant Raymond FitzMaurice was placed in charge of the development of instruments for use in the navy's seaplanes and early the following year he carried out tests in a Short seaplane fitted with a transmitter and receiver, piloted by J. Babbington; communication was established over a distance of forty-five miles with both ground stations and ships. In the same year Lieutenant James, in a BE aeroplane and using a receiver in which the output was amplified by a Brown relay and which was screened from magneto interference, succeeded in receiving signals with the engine running at full power.

In the naval manoeuvres of the same year Lieutenant Fitz-Maurice and Commander Samson, in a Short seaplane based on Great Yarmouth, were making a reconnaissance over the sea when their engine failed and they were forced to land on the sea. Because of the signals they had sent out the ship *Hermes* was able to locate them, which must have been the first instance of wireless being instrumental in the rescue of a ditched aircraft.

Some idea of the progress made in this field can be gained from the fact that, when presenting the 1913 Naval Estimates to Parliament, Mr Winston Churchill, First Lord of the Admiralty, announced that of the forty seaplanes then in service, six had been fitted with wireless telegraphy giving them a range of sixty miles. During 1913 too there was further development of the use of wireless telegraphy in airships. The army airships *Delta* and *Eta* took part in the army manoeuvres of 1913 and succeeded in communicating with each other at a distance of a hundred miles.

Germany was also developing airship wireless telegraphy. Her Zeppelin LI was one of those fitted and she used it to check her position and receive time signals. While over the Heligoland Bight she was caught in a storm and disabled; she sent out a

---

* Some technical notes on the apparatus used in aircraft at this time will be found in the appendix.

distress signal, probably the first from an airship, but help arrived too late to save her crew.

The Yearbook of Wireless Telegraphy and Telephone for 1913 contains a very interesting table giving details of the sets which the Marconi Company had developed at that time for military purposes.

TABLE 4

Portable Military Stations, 1913

Type	Description	Max Power kW	Max range m	No of horses required for transport	Weight lb	Height of mast ft
K	Knapsack	·04	12	By hand	86	30
A	Pack	·50	50	4	804	30
A1	Pack (special)	·50	80	5	964	54
C	Handcart	·50	50	1	1,157	30
F	Cart	1·50	250	4 or 8	5,024	70
L	Aeroplane	·04	12		50	
L1	Aeroplane	·50	80		200	
M	Dirigible	1·50	300		500	
H1	Motor car	·50	150	20–5hp car	4,043	70
H2	Motor car	3·00	300	20–5hp car	7,800	70

In view of the experience which the services had with some of the sets under actual war conditions, the column headed 'weight' is of particular interest. Two of the sets are described as for use in aeroplanes, for which 200lb would have been very heavy. One would think too that the market in Great Britain for the dirigible set weighing nearly a quarter of a ton must have been very limited; one wonders how many of them were sold to the Luftwaffe by the company's subsidiary in Germany and whether any of them returned to England as part of their Zeppelins' equipment.

7
World War 1 at Home and at Sea

The position of wireless telegraphy when war broke out in 1914 may be summed up briefly: although great technological progress had been made, and although it had been shown conclusively to be of great value in numerous fields and had been instrumental in saving many lives at sea, it had as yet made no real impact on the lives of the great majority of people. Even at sea it was still an optional extra on many ships and only in the Royal Navy and aboard large liners was it regarded as an absolute necessity. A considerable part of the British merchant fleet was still without it. Its use in the air was still experimental, and the possibility of it becoming an important factor in the conduct of a war had been given much less thought than it should have been. At the same time the development of radio telephony had not progressed very far, and only in the United States do more than a very few people seem to have envisaged its potentiality as a source of entertainment.

When war came, not only were the supplies of equipment inadequate or unsuitable but there were still some people in high places whose attitude towards wireless telegraphy revealed ignorance and lack of imagination; and nowhere was this more apparent than in the regulations promulgated under the Defence of the Realm Act regarding the use of wireless telegraphy by civilians. Anyone who has now reached early middle-age will remember the part played by 'the wireless' in everyday life during World War II. Not only was it a source of up-to-the-minute news but the government regarded it as an essential means of communication with the public and of maintaining morale. 'Wireless' was in this case sound broadcasting, and it does not seem to have occurred to anyone that the possession of a receiver capable of receiving signals from anywhere in enemy territory made a man a bad security risk.

In 1914, however, the official attitude towards the public vis-à-vis wireless was quite different. On the outbreak of war its use by civilians was prohibited. The all-embracing nature of this prohibition can be seen from the following paragraph in the regulations under the Defence of the Realm Act:

> No person without the permission in writing of the Postmaster General shall buy, sell, or have in his possession any apparatus for sending or receiving messages by wireless telegraphy, nor any apparatus intended to be used as a component part of such apparatus.

Furthermore, anyone possessing any kind of wireless apparatus was required to declare it immediately.

This amounted to a complete shut-down of all wireless stations except for official use, and the absolute prohibition of all amateur activity. The only exception was the transatlantic service.

The way the regulations were enforced seems hardly credible today. One schoolboy known to the author possessed a 'receiving station' which consisted of a very crude crystal receiver hitched to an aerial hung from a short wooden mast. He duly declared this, and a few days later an army officer called to inspect it. The officer directed that the receiver should be placed in a drawer, which he proceeded to seal. Nor was this the end of it. For more than a year the officer called once a month to make sure that the seal was intact. Eventually it must have occurred to someone that the whole performance was an appalling waste of manpower, for one day the crystal set was taken away to be watched over in some central place of security and was never seen again.

Anyone caught in possession of an undeclared set, components, or even just an aerial was treated as a dangerous enemy agent. There were numerous prosecutions of people found to possess undeclared crystal sets, and the penalties inflicted by magistrates, who probably had little idea of the potentialities or limitations of such instruments, were severe. On 19 November 1914 at Blyth, Northumberland, a young man was particularly harshly dealt with. Someone visiting his house had seen that he had a room which appeared to be full of electrical equipment. The visitor went to the police and the room was searched. It was found to contain a small wireless transmitter. In court it was admitted by a Post Office engineer appearing for the prosecution that the transmitter had a range of no more than five miles under favourable conditions (say, just enough to contact a submarine that had surfaced in full view of the shore). The young man in fact made a hobby of electrical experimentation and there was no evidence that he had ever been guilty of anything more than failing to register his equipment. Nevertheless the magistrate declared that he was potentially a serious danger to the country and sentenced him to nine months' imprisonment.

The official attitude towards students being trained for the wireless service at sea was even more incredible. All these young men were vetted for security before being allowed to start training, yet the training schools were not allowed to connect their receivers to outside aerials. The result was that the students, on completing the course and now holding Post Office certificates showing them to be proficient to take charge of the wireless station on any ship, had never heard a live wireless signal. Many were sent to sea alone on their first voyage, and this despite the fact that the ship might be involved in enemy action within an hour or two of sailing.

No doubt this attitude arose from the fact that during the early months of the war the country suffered from a severe attack of spy mania. For several years before the war the expression 'German spy' was frequently heard and the public had been

Page 103 (*top*) A replica of a wireless cabin of 1920 at the Marconi Marine Jubilee Exhibition, Baltic Exchange, 1950. The equipment includes (from left to right): type 11F DF; buzzer and key (below the call sign plaque); switchboard type 104A; receiver amplifier type 91; receiver type 31C; receiver AT1; emergency break; 1½RW QG transmitter and switchboard type 95; (*bottom*) a Marconi type 12A direction finder (about 1920)

Page 104 (*top*) The world's first wireless telephony news service was inaugurated on 23 February 1920 at the Chelmsford Works with this 6RW transmitter, using a wavelength of 2,500m. At the microphone is W. T. Ditcham; (*bottom*) a Marconi mobile transmitter/receiver, complete with power generator, driven by the engine of a Douglas motorcycle

conditioned to regard every German waiter and bandsman as a member of Kaiser Wilhelm's espionage service. No doubt such a service did exist, but having regard to the technical position at the time it is highly unlikely that anyone could for long operate a transmitter capable of reaching the Fatherland or even a submarine well out to sea without being discovered, for such a transmitter would have been extremely noisy and would have needed an elevated aerial.

A more rational restriction was that which prohibited the use of code or cypher in telegrams transmitted across the Atlantic by wireless. This was a serious handicap to the service as the need to write commercial messages in plain language made it more expensive than the cables. The difficulty was eventually overcome by permission being given for the use of a limited agreed code.*

As there was no broadcasting, wireless could not yet be used for propaganda direct to civil populations abroad. Consequently Germany had no Lord Haw-Haw. However, she did have a high-power station at Nauen and this was used to put out an endless stream of news bulletins. Doubtless these were mainly intended for neutral consumption, but if any news was quoted from them by the British press it was presented in such a way as to imply that it was a comic distortion of the truth. The propagandists had to wait until the spoken word could be heard in every home before wireless could be used as a weapon of psychological warfare on a large scale.

While at home the witch-hunt for 'traitors within the gate' was in full cry, abroad energetic steps were being taken to destroy Germany's means of communication with friends beyond her borders. In 1914 she had a number of colonies in Africa and the Pacific, and as we have seen unlike Great Britain she had not allowed politics to delay the construction of a chain of high-power wireless stations. Unfortunately for her, however, they

* *Cypher* comprises groups made up of any combination of letters or figures and is chiefly used for official purposes. *Code* is made up of combinations of letters forming pronounceable syllables and is mainly used in commerce.

G

were mostly sited in vulnerable positions. The allies had no doubts about their value to Germany and very soon after the outbreak of war they set about eliminating them. The first to go were the stations at Dar-es-Salaam, East Africa and at Yap in the Caroline Islands, both of which were destroyed on 12 August 1914. On 28 August the Germans themselves destroyed the big station at Tanina in Togoland, and on the same day the American government closed down the German-owned high-power station at Tuckerton, New Jersey. A few days later the German stations in Samoa and at Nauru in the Marshall Islands were captured by the Australians, and on 12 September Herbertshohe in Neu Pommern was destroyed.

When in July 1914 it became clear that war was imminent, of the three fighting services only the navy was adequately equipped with well-tested and efficient wireless apparatus manned by a sufficiency of fully trained and experienced personnel. British warships had been among the first vessels to be fitted with wireless telegraphy and the navy had its own shore stations at home and at numerous strategic points overseas. As a result, the history of naval wireless during World War I was not so much an account of the adaptation of apparatus and methods to the needs of the service—that had already been done—but of the effect of it on operations in the first naval war since its invention.

In 1914 Britain possessed no wireless stations with worldwide range; nevertheless, by using a combination of British-owned cables and naval wireless transmitters at overseas stations it was for the first time possible, on an occasion of major importance, for the Admiralty to communicate an urgent order to all or nearly all ships and stations simultaneously. Winston S. Churchill in his *World Crisis 1911–18* describes how this facility was employed to broadcast two fateful messages. The first went out at 1730hr on 3 August 1914 and read:

> Admiralty to all ships—Urgent message. The war telegram will be issued at midnight authorising you to commence hostilities

against Germany but in view of our ultimatum they may decide to open fire at any moment. You must be ready for this.

At 2300hr GMT (midnight German time) the second message was sent to all naval ships and establishments,

Commence hostilities against Germany.

The word 'broadcast' originally described a method of sowing seeds, and it is doubtful whether it was used generally or even at all at this date in connection with wireless. Certainly general calls were made and messages were transmitted simultaneously from one station to a number of others in the vicinity. Generally speaking, however, the spectacular triumphs of wireless so far had been concerned with the ability to communicate between a fixed and a moving point. This operation demonstrated its use in communicating information from a central headquarters simultaneously to a large number of reception points.

The power to send orders to distant ships instantaneously was exploited with great effect by the British Admiralty. However, it was a power that was also possessed by the Germans. This meant that in effect each side was able to eavesdrop on the other's orders. As one would expect, the orders were in cypher and unless the enemy's cypher could be cracked listening-in to the other side's transmissions was not a very productive exercise. Here, as the result of an incident which resembled something out of a yarn by an imaginative thriller writer rather than an actual event, the British Admiralty had a tremendous advantage over the Germans.

It would be impossible to improve on the account of this incident to be found in *World Crisis 1911-18*.

At the beginning of September 1914 the German light cruiser *Magdeburg* was wrecked in the Baltic. The body of a German under-officer was picked up by the Russians a few hours later, and clasped to his bosom in arms rigid in death were the cypher and signal books of the German navy, and the minutely squared maps of the North Sea and Heligoland Bight.

On 6 September 1914 the Russian naval attaché in London called on Winston Churchill, First Lord of the Admiralty, and informed him of this, adding that the Russian Admiralty felt that these cyphers and charts should be transferred to Britain, the leading naval power. Churchill despatched a ship to the Russian port of Alexandrov to fetch the officers in charge of them, and early in October the precious documents were handed over at the Admiralty to Churchill and Prince Louis of Battenberg, the First Sea Lord.

The impact which this coup had on the conduct of the war cannot be exaggerated. The British blockading forces were stationed at Scapa Flow and the Straits of Dover. Between them all the east coast ports and towns lay open to attack by German raiders. At first sight this may seem to have been a strange disposition of forces, but there was no alternative. It was necessary to keep the blockading forces in harbour at strategic points until they were needed. To have maintained at sea a force strong enough to repel any possible attack would have placed too great a strain on men and ships. However, there was now no need to. By monitoring all German naval transmissions and deciphering them with the aid of the cypher books from the *Magdeburg* it was possible to predict a raid and have a suitable force waiting for the enemy squadron.

One would have thought that the risk of their cyphers being cracked would have been obvious to the German commanders, and that they would have maintained wireless silence. Admiral Scheer deals with this point in the German official history:

> In the case of large forces whose units are stationed far apart, communication between them is essential and absolute cessation of all wireless intercourse would be fatal to any enterprise.

This may have been true; nevertheless one cannot help asking why, given that it was necessary for his ships to communicate while in the Heligoland Bight, they used so much power that they could be heard on the other side of the North Sea. When the convoy system was organised many British merchant ships were fitted with 'convoy sets' which were in fact buzzers, for communication between ships of the convoy, but their power was so small that their range hardly extended to the horizon. But even this practice was not universal. The author sailed in a number of convoys in the North Sea, Eastern Atlantic, and Mediterranean, some numbering as many as fifty ships, in which there was no wireless transmission whatever.

The British Navy seems to have dealt with the question of telltale wireless signals much more effectively than did the German Navy. This was illustrated by the events which led up to the Battle of the Falkland Islands. When war broke out the Germans had a formidable squadron, which included the powerful armoured cruisers *Scharnhorst* and *Gneisnau* under the command of Admiral von Spee, stationed in the Pacific. This force offered a serious threat to allied shipping not only in the Pacific but also in the Atlantic. The only force in the Pacific which could be brought against it was a light cruiser squadron under Admiral Craddock. That this was much inferior in strength was sadly demonstrated when the two squadrons met off Coronel on the coast of Chile and Admiral Craddock was defeated with the loss of his two best ships, the *Good Hope* and *Monmouth*.

After this disaster the Admiralty decided to despatch an overwhelming force to deal with von Spee. On 11 November 1914 the battlecruisers *Inflexible* and *Invincible* sailed from Devonport, to be joined later by six smaller cruisers. The combined force arrived at Port Stanley, Falkland Islands, without having once used its wireless, and so well had the secret been kept that it was not until von Spee was approaching Port Stanley the following day that the sight of tripod masts over a low spit of land gave him his first indication that the British Navy had battlecruisers in that part of the world.

Eventually it was noticed by the Germans that every time they sent out a force to raid the coast of Britain it was met by a British squadron that was evidently expecting it, and it became clear that the Royal Navy was able to decipher their wireless messages. They accordingly changed their cyphers and the *Magdeburg* books were no longer of value. By this time, though, the Royal Navy had another weapon—the wireless direction finder.

This instrument had been brought to a state, if not of perfection at least of practical use, shortly before the war, and the navy started to erect DF stations immediately after the commencement of hostilities. These were able not only to determine the position of any German ship that left harbour and was indiscreet enough to use its wireless, which we know it was very likely to do, but by making a series of observations was able to track her course across the North Sea and so give warning of the point on the coast of Britain which was likely to be attacked.

The technology of DF was not secret and there seems to be no reason why the Germans should not have been equally well equipped to track any British ships that had been free with its wireless, but in fact they were not.

Admiral Scheer wrote:

> The English received news through their 'directional stations' which they already had in use, but which were only introduced by us at a much later date. In possessing them the English had a very great advantage in the conduct of the war, as they were able to obtain quite accurately the position of the enemy as soon as any wireless signals were sent by him.

By far the most fateful misuse of wireless by the German Navy was that which led to the Battle of Jutland (see Fig 6). At this date the British Navy had DF stations sited at Lowestoft, Flamborough, and Aberdeen, and on 30 May 1916 they noticed considerable wireless activity on the part of the German naval wireless telegraphy control vessel *Bayern* lying at Wilhelmshaven. It was then noted that the bearing of this ship had changed $1 \cdot 5°$, showing that she had moved nearer the sea. This was reported to Admiral of the Fleet Sir Henry Jackson, First Sea Lord, who took it as an indication that the German high seas fleet was

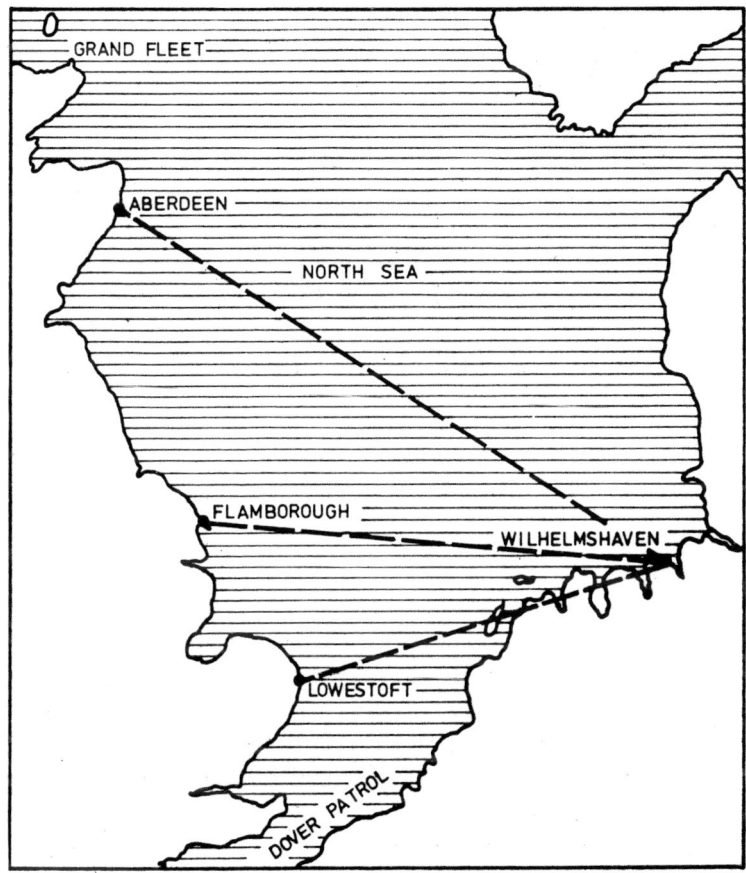

Fig 6 Position of blockading fleets and of DF stations monitoring movements of German fleet

about to put to sea. Accordingly he ordered the British grand Fleet to put out from Scapa Flow with the result that the battle of Jutland took place next day.

This was the most momentous use of wireless DF to date. One of the most remarkable aspects of the affair was that, remembering that DF had been in practical use only for some five years and for much of that time on a small scale, the First Lord of the Admiralty had taken such a tremendous decision on the basis of a change of bearing of only 1·5°.*

Once, then, an efficient DF system had been developed, no ship or aircraft could use its wireless without risk of betraying not only its presence but its precise position. It is of course not meant to suggest that the British Navy did not use its wireless during the war; it was simply that it was much more careful with it than were the Germans. One important use to which it was put was in enabling submarines to keep in touch with their depot ships. A drill for this purpose was practised which may be summarised as follows:

Surface
Up mast
Brief transmission
Down mast
Submerge

The German U-boats, which had to operate a long way from base and had no depot ships, used their wireless a lot more than did the British submarines. Royal Navy anti-submarine patrols were able to take considerable advantage of this and were able to track a German submarine by DF as it proceeded on its course and so forecast its movements. The work of tracking and destroying submarines was carried out in close co-operation with the Royal Naval Air Service and was directed by Admiral of the Fleet Sir Arthur K. Wilson, VC.

In one area the use of wireless by the navy was less successful. During the bombardment by the fleet of the Dardanelles attempts were made to use RNAS planes as spotters for the gunners, in the same way as the RFC was co-operating so successfully with the

* The Bellini-Tosi direction finder was invented in 1907 but at first was limited in range because of the lack of a sufficiently sensitive receiver. It was not until valves were introduced that it became of much use. The receivers used in connection with the DF stations mentioned above comprised one HF stage, a balanced crystal detector, and two LF stages. The valves were 'soft' valves, very much less efficient than the 'hard' valves introduced a little later.

artillery on the Western Front. The naval gunnery officers were not used to this kind of spotting; they showed little interest in it and anyway there had been insufficient practice, nor was the wireless equipment satisfactory, the aircraft having been fitted with the new Sterling sets which were still in the teething-trouble stage. Furthermore, the arrangements for charging batteries were inadequate and the results obtained were very disappointing.*

The merchant marine, like the Royal Navy, entered the war with a fully operational and—so far as the ships already fitted were concerned—adequately equipped wireless service. Its installations had been designed for the purpose and had been in use for a number of years, and it was not found necessary to modify them for war purposes.

There was, however, need for considerable expansion. Just before the outbreak of hostilities regulations had come into force requiring all ships carrying fifty or more persons to be so equipped and the work was put in hand as quickly as possible. However, there were shortages of materials and manpower to contend with and the war was well advanced before the task was completed. Strange to relate, the attitude of many captains and officers to the fitting of their ships was often by no means welcoming. When one remembers that all the time a ship was on the high seas she was at risk of sudden attack and sinking, and that if she was unaccompanied when sunk her crew might find themselves adrift in open boats with nobody aware of their fate, much less of their position, one would have thought that they would not only have welcomed the fitting of their ships with wireless but would have demanded it. Nevertheless, the wireless officer appointed to a newly fitted ship often found that he was far from welcome. The captain and deck officers tended to regard him as

* Not only did the Royal Navy enter the war well equipped with wireless installations, but it continued to expand its wireless service and keep it up to date. By 1915 it had erected fourteen new shore stations sited at strategic points; eleven were of 5kW, two of 20kW, and there was one 120kW rotary spark station. Arc CW transmitters were in use by 1917 and in 1918 valve transmitters were introduced.

an intruder and professed to have a poor opinion of the value of wireless, whilst the engineers complained at having to start up the dynamo for daily tests. (At that time tramps usually had only oil lighting and when wireless was fitted a dynamo had to be installed for its power supply.)

This prejudiced attitude of some of the officers is well illustrated by an experience which the author had while serving on an ancient tramp during the last year of the war. The ship was proceeding from Genoa to Marseilles. About midnight, when it was very dark, the gunner on duty sighted a vessel which he took to be a submarine passing close by in the opposite direction. Whatever she was, she turned and proceeded to follow the tramp. The elderly (dug-out) officer of the watch did not think she was a submarine while the captain was uncertain and hesitated to give the order to fire. Meanwhile the wireless officer was blissfully taking down a long news bulletin. When he had finished he went up to the bridge to arrange to be called for his watch. By this time the chase had been going on for half an hour, but neither the captain nor the officer of the watch had been sufficiently wireless-minded to think of having the wireless officer called, much less of ordering him to send out the brief code message: 'I am being attacked by a submarine'.

The wireless officer promptly asked the captain if he should have the dynamo started up but even at this stage the captain was in no hurry to use his new facilities, replying that there was no need to do so yet. However, the wireless officer thought differently. He had no battery-operated emergency transmitter, and so he knew that if the ship should be torpedoed he would have no chance of getting the dynamo started. Accordingly he took it upon himself to disobey orders. And it was fortunate that he did so, for very shortly afterwards he was ordered to send out an SOS.

It is interesting to compare the absolutely minimal equipment supplied on tramps at that period with the elaborate instrumentation to be found on quite humble ships today. This is illustrated by another incident which took place aboard the same ship on a voyage to Archangel in Arctic Russia, and back.

While in the Russian port the temperature had dropped to −40° F. Despite this the steam heaters were turned off all night as an economy (tramp-ship style). The steam heaters in the accommodation promptly froze and some of them, when thawed out with a blowlamp next morning, cracked, allowing steam to escape from them. Among these was the one in the chartroom. Some of the steam found its way into the chronometer, where it promptly froze again. Accurate time-keeping is essential for navigation, but now there was none, for the owners had supplied only one chronometer instead of the safe number of three.

So long as the ship was in convoy this was not serious, but when a few days later off the north of Scotland she became separated from the convoy it was another matter and if it had not been for the despised wireless the captain might have found it very difficult to find his destination. However, the wireless officer was able to receive time signals from the Eiffel Tower, then the only available source. He noted them on his own watch and the captain then transferred them to his. It was not the textbook procedure but it proved adequate for getting the ship into port—and also to show that wireless was worth having.

Until practically the end of the conflict the outcome of the war against allied shipping was in the balance. There can be no question but that wireless saved shipping, by helping the naval forces to locate and destroy submarines and by saving the lives of many seamen so that they lived to man new ships. The Germans, of course, also had wireless which technically was just as efficient but in the submarine war it was relatively less important to them. And in the naval war they used it with such ineptitude that it was really a handicap rather than a help. In summing up, therefore, it can be fairly said that wireless played a major role in winning the victory at sea for the allies.

8

World War 1 on Land and in the Air

If during the prewar years the British Army had had a Lord Fisher it would certainly have entered the conflict much more adequately supplied with wireless equipment and with personnel much better trained to handle it than in fact was the case. As it was, it possessed only ten sets which, as recorded earlier, had been supplied for use with the cavalry—ten sets that is which were soon found to be unsuitable for their purpose. And when one considers the advances which had been made in other applications of wireless, it seems reasonable to suggest that much more could have been done to equip the army with a really efficient wireless service. The Marconi Company had developed a whole range of equipment for military purposes and they could hardly be blamed for designing it with the Boer War in mind. With more interest and enthusiasm in high places there would have been more accurate knowledge of what was needed.

During the first few weeks of the war, that is to say during the period of the retreat from Mons, army wireless telegraphy was a complete failure. This was due to a number of causes, the first being the inexperience and insufficient training of personnel. Next to this was the clumsiness of the equipment, such as, for instance, the cart set, which was mounted on two limbers, weighed in all two tons, and required six horses to draw it; it also took twenty minutes to erect the mast and aerial and get the station operational. Thirdly, there was selectivity, or rather the lack of it, something that was to plague the wireless telegraphy service on the Western Front for some time to come.

Although there were then only ten stations, this third problem was serious and in an attempt to overcome it what was called the 'period system' was devised. Under it only one station was allowed to transmit in any ten-minute period. This led to a message from station A to station B being dealt with in something like the following manner:

Station A   Transmits message
Station B   Transmits query
Station A   Correction
Station B   Acknowledgement

This meant that even a short message could take up to forty minutes to transmit and that all other stations would have to remain silent during that period. It also meant that if the commanding officer of a force on the move wished to send an urgent message the timetable would work out like this:

Set up station   20 minutes
Send message   40 minutes
Dismantle station   20 minutes

A total of eighty minutes, plus the time wasted if the operator found that the air was already occupied by another station. It is not surprising that some officers viewed wireless telegraphy with disfavour.

Lieutenant-Colonel Chenevix-Trench ascribes much of the trouble in the field not so much to rapid movement as to too frequent movement; he says that the sudden and frequent orders made operations difficult. He goes on to comment that the tactics of communication had not been sufficiently studied, and that such study should have been at staff level and not left to the field officers.

It is perhaps some consolation to reflect that German military wireless was equally unsatisfactory during the early stages of the war. Anthony Farrar-Hockley in *Death of an Army* describes the deplorable state of affairs behind the German lines. The German GHQ at Coblenz relied on wireless telegraphy to keep in touch with its right flank and centre but its service was no better than the British. Its equipment was primitive and its procedures cumbersome. All messages had first to be enciphered, then transmitted by hand-operated morse, and finally deciphered, so that it took several hours for a message of any length to reach its addressee, sometimes even days. Nor were operations helped by the jamming service provided by the Eiffel Tower and increasing frustration led to parts of secret messages being sent in plain language. The general communications situation was made still worse by inefficient telephones and the cutting of telephone and telegraph wires by the inhabitants of occupied territories.

Nor did the Russians get much help from wireless during this early period. Quite the contrary, as they repeated the mistakes they had made in the war with Japan, once more sending secret information in plain language. In fact the Germans found this so useful that for some time they were reluctant to develop their own wireless service, being afraid that it would lead to similar leakages of information, an attitude shared to some extent, as we have seen, by all the allied commanders.

With the stabilisation of the front in the autumn of 1914 the cavalry ceased to function as such and fought alongside the infantry in the trenches. In forward positions it was found that aerials were much too conspicuous, so wireless companies were transferred to interception work. But the use of line telegraphy in the front line was far from satisfactory, being seriously hampered by the constant cutting of wires by shellfire. Major-General R. F. H. Nalder CBE, in his *The Royal Corps of Signals*, tells us that this led in the summer of 1915 to officers being appointed to reconsider the possibility of using wireless telegraphy. What was required was a reliable short-range portable set, but none was available, the nearest being the Marconi $\frac{1}{2}$kW pack set which was too heavy and clumsy. However, by August a design had been worked out for a new set and an order was placed for a hundred. This set became known as the British field wireless set. The smallness of the number ordered indicates the limited confidence placed in wireless telegraphy by the army at that time and when the sets were delivered, which was not until nearly a year later, the number was found to be quite inadequate. The first use of these sets was at the Battle of Loos. Although results were encouraging, the staff still looked on them with suspicion. One objection to them was that they could be located by the German DF, though this does not seem to have happened.

Trials were also made with telephones in forward areas and it soon became clear that conversations were being intercepted. Regulations relating to 'telephone discipline' were not observed as carefully as they should have been, particularly by senior officers, and thousands of casualties were attributed to the consequent leakage of information. The trouble was eventually traced to the use of an earth return and when single telephone cables were replaced by twisted twin cables this cause of trouble was eliminated.

There was great need for wireless communication in forward positions but this was hampered by numerous difficulties. Unless great care was taken in their siting, masts could become targets for enemy gunners; all equipment was very vulnerable to damage by shellfire; accumulators for operating the transmitters had to be transported by hand under very difficult conditions and they often arrived minus much of their acid; the gear supplied was clumsy; there was a severe shortage of skilled operators; and organisation was poor. The BF set had a range of 4,000yd with an aerial 60yd long supported by 12ft masts though with a better aerial it could reach 10,000yd. These figures assume the use of crystal detectors, with which they were designed to work. At the time of which we are writing Dr W. H. Eccles was working on the development of the valve and using a valve receiver in London he was able to pick up signals from trench sets being used on the Western Front. The situation was much improved by the introduction in

1917 of the first CW sets. With 10yd aerials only 2–3ft high they had a range of 6,000yd.

Major-General Nalder in *The Royal Corps of Signals* discusses the difficulties with which the army wireless service had to contend on the Western Front:

> Firstly commanders and staff had little faith in wireless. In consequence it tended to be cold-shouldered and divorced from the normal chain of command. Secondly the security problem masked the issue. Thirdly the War Office organisation was not best adapted to directing development. Even if the Director of Signals had been more persistent in pressing his demands, it is doubtful whether under that organisation the resources of the UK could have stepped up provision to any marked extent. Some of the most acute problems arose from the terrible conditions of trench warfare, battlefields for which there was no precedent on a comparable scale. The struggle to find a way out of the impasse was shared by the whole of the armies, allied and enemy alike.

Reviewing the performance of wireless telegraphy on the Western Front he writes:

> While the telephone service was inevitably dependent on lines, the telegraph service need not have been to the extent that it actually was. The wireless services were neglected and were not used sufficiently to make them efficient in an emergency.

So far we have been considering the use of wireless telegraphy for maintaining communications in forward areas and the picture we have had to draw is not a very happy one. Wireless telegraphy was, however, used for another purpose, one for which there was no practical alternative, that is in connection with the use of aircraft as spotters for the artillery. In this it was outstandingly successful and for a consideration of it we must turn our attention to the activities of the Royal Flying Corps.

When war broke out the RFC had been in existence for just over two years and was under the control of the War Office, there being no Air Ministry at that date. Its personnel retained military ranks and it was, therefore, an arm of the army.

It had been envisaged that the role of the aeroplane in war would be to carry out reconnaissances behind the enemy line—to learn, in fact, what was happening on the other side of the hill—and to do this it might have to penetrate deep into enemy territory; accordingly the sets developed for use in aircraft were designed to have the maximum possible range. Table 4 (p 99) shows that just before the war there were two sets available for this purpose. One had a power of 40W and a range of 15 miles, weighing 50lb. A larger one was rated at 0·5kW, had a range of 80 miles and weighed 200lb and was therefore capable of filling the requirements as to range but was much too heavy and bulky for the aircraft of that date.

However, in the early stages of the war on the Western Front aircraft were used almost exclusively for the purpose already mentioned—spotting for the artillery. It was an established maxim in military circles that artillery was no better than its forward reconnaissance and it was soon evident even to the most conservative of artillery officers that there was no better place for their spotters than up in an aeroplane. But the spotter must be able to communicate his information to the gunners with a minimum of delay and at that date there was no established means of communication between an aeroplane in flight and the ground.

A variety of means were tried which included the dropping of Véry lights and the display of flags as well as wireless telegraphy. Some machines were even fitted with klaxons, the plane being turned so that the klaxon faced the guns and the message then being boomed out in morse. When one thinks of the noise that must have been going on, with guns firing and shells exploding, not to mention the difficulty of sending readable morse under such conditions, one wonders whether messages were ever received correctly. Of these methods by far the most successful was wireless telegraphy though it still needed considerable improvement to make it really satisfactory.

Lieutenant (very soon Lieutenant-Colonel) Swain Lewis was the first to use wireless telegraphy for this purpose. It was very soon found that the sets fitted were quite unsuitable as apart from their excessive weight they occupied the whole of the observer's seat in a two-seater plane; they were also much too powerful as the spotter planes needed to fly only a very short distance from the ground station. And the sets then in use created much too much inteference between the aircraft at work close to each other.

If, then, the aeroplane could not carry an observer, the pilot had not only to control the machine but had also to observe where shells fell and transmit the information to the ground—quite a feat. In Walter Raleigh's *The War in the Air* there is this example of a series of messages received from a spotter plane on 24 September 1914:

4.2 p.m.	A very little short. Fire, fire.
4.4	Fire again. Fire again.
4.12	A little short. Line O.K.
4.15	Over, over and a little left.
4.20	You are just between two batteries. Search two hundred yards either side your last. Range O.K.
4.22	You have them.
4.26	Hit, hit, hit.
4.32	About 50 yds short and to right.
4.37	You are in the middle of three batteries in action. Search all round within 300 yds of your last shot and you have them.
4.42	I am coming home now.

Such a procedure was not only cumbersome, its accuracy left much to be desired, and Lewis set about adding a grid of lines to his army maps, enabling references to be read off quickly and correctly. These maps were seen by a staff officer at RFC HQ who was much impressed, and as a result Lewis's system was adopted for all army and RFC maps.

As the war progressed wireless telegraphy grew in importance and the wireless unit was expanded to become No 9 Sqn RFC. It was equipped with two machines fitted with wireless telegraphy and there was an arrangement by which any battery wanting a 'wireless aeroplane'—the description then in use—had one seconded to it from the squadron. As there were only two of them and the calls on the services were very great, their pilots were heavily overworked. Eventually the squadron was returned to England, to become the RFC's first wireless training unit, and its duties in France were taken over by other squadrons. As soon as possible every battery had a spotting-plane attached to it and these soon came to be regarded as essential.

It was quickly discovered that wireless interference between these spotting-planes was of serious proportions, so much so that a point was reached when the number of guns which could be used along a given length of front was limited by the number of planes' transmitters being used in the area, and this of course was in turn controlled by the selectivity of the receivers on the ground. Two means were employed to cope with this difficulty. One of them was the introduction of a device known as the 'clapper break', which made it possible to vary the pitch of the note transmitted, this in turn enabling the ground operators to distinguish between the required signal and one jamming it. The second was the introduction of the Sterling set. This was manufactured by the Sterling Telephone Co Ltd to the designs of Lieutenant Leroy, an RNVR officer in the RNAS. It was a good deal less powerful than the sets previously in use and caused much less interference. It was also compact and light, weighing only 20lb. With these improvements it became possible to double the number of guns which could be used. However, up to the Battle of the Somme in 1916 it was still only possible to use one plane per 2,000yd of front, though later improvements in selectivity enabled the density to be doubled.

The elaborate air-ground wireless telegraphy organisation built up during the static period of the war in France seems to have broken down when the armies started to move. In the great battle in March 1918, when the British Army was retreating, the officer commanding No 8 Sqn recorded:

> I was unable to find out the position of any of our batteries, so I sent out my wireless officer with as much spare gear as possible,

so that he might be able to get battery stations working where portions of the equipment had been lost. He found several batteries, but not one with a mast working. As soon as the retreat started all idea of co-operating with the aeroplanes seems to have been abandoned. Many batteries had simply thrown their equipment away, others had retained the instruments only. Under the circumstances little use was made of the calls that were sent down, the answering of which was probably the only hope the artillery had of hindering the German advance.

On the other side of the line the Germans were experiencing similar trouble, in their haste to advance having abandoned much of their wireless equipment. And in their case the situation was aggravated by the fact that just before the battle began the Luftwaffe operators manning the ground stations had been replaced by others supplied by the artillery who were not familiar with the instruments and procedure. It is easy to see that if they had succeeded in maintaining an effective air-artillery spotting organisation, at a time when the British system had broken down, the outcome of the battle might have been very different. But then again, when the tide had turned and the British Army was advancing on the Hindenburg Line, it was reported that units were often slow to establish stations as they moved forward so that opportunities of inflicting damage on the enemy were lost.

These accounts show clearly the tremendous impact which wireless had made on the conduct of the war in France, despite its poor performance as a means of communication between units in forward positions. A few figures will illustrate the rapid growth of air-ground wireless telegraphy. At the Battle of the Somme in 1916, 316 planes were fitted and there were 542 ground stations. When hostilities ceased two and a half years later 600 British aircraft were equipped for air-to-ground wireless telegraphy and there were about a thousand ground stations, with a range of fifteen miles. Long-range bombers were by then equipped with cw transmitters.

The first satisfactory radio-telephone sets were introduced in 1915; these had a specially designed microphone for use near the engine. However, an unofficial experiment with the use of RT in aircraft had been made a little earlier. Basil Collier, in his *Leader of the Few*, a biography of Air Chief Marshal Lord Dowding, recounts that in 1914 the then Captain Dowding, in command of No 9 Sqn RFC, with the assistance of C. E. Prince, who before the war had had professional wireless experience, rigged up a wireless telephone transmitter and fitted it to a Maurice Farman biplane. This was taken up and Captain Dowding was able to claim to be the first man to receive telephony from the air in England, perhaps anywhere. However, when the War Office heard about it they declared that radio telephony with aircraft was impracticable and decreed that the experiments must cease.

For a large part of the war transmission was one-way, because the noise of the engine made reception in the air difficult. However, as time went on aircraft, particularly those used by the RNAS, were given receivers, usually consisting of a crystal detector plus a one-valve amplifier. The crystal detectors were of the cat's-whisker type and it must have been difficult to keep them critically adjusted in a vibrating aircraft. Later, balanced carborundum crystals were installed and these were much more stable.

The appearance of tanks on the battlefield brought new problems for the RFC. It was soon found that close air–tank co-operation was essential and from July 1918 No 8 Sqn was attached to the Tank Corps and experiments were carried out to determine the best means of communication. The tanks being mobile, it was impossible to use fixed ground stations, so communication had to be direct between the air observer and the tank. RT, which by then was coming into use, was tried but it was found that even under the most favourable conditions signals could only be heard in the tank when the aircraft was not more than a quarter of a mile distant and less than 600ft above ground. Wireless telegraphy was then tried and found to be successful up to 9,000yd distant and a height of 2,000ft; it was, however, necessary for the tank to stop and erect an outside aerial before it could be used, which was not very desirable. One of the most urgent needs was to communicate with halted tanks to determine whether they were disabled or had halted for some other reason. Pending the fitting of all tanks with wireless telegraphy smoke and signal bombs were used.

Army wireless made a much better showing in the subsidiary theatres of war, very largely because there was often no other means of communication and so it was viewed more favourably. They were not static campaigns bogged down in trenches but often involved considerable movement which meant that the sets that were available when war broke out were much more suitable for them. Also, there was little or no risk of interception by the enemy. From Chenevix-Trench we learn that wireless telegraphy was particularly successful in Palestine. There operations mainly took the form of a cavalry campaign and often wireless telegraphy was the only means of communication between 4th Cavalry Division HQ and Desert Corps HQ. The 4th Cavalry Division had four pack sets and the Desert Corps two pack sets in Ford cars, a wagon set, and a prewar car set. The longest distance covered was 140 miles. He notes that in addition to their wireless telegraphy duties, the hard-pressed operators had to care for their horses. In this campaign it was found that due to the heaviness of the traffic which they had to handle and the consequent adequate practice which this involved, a high standard of efficiency was maintained.

By the middle of 1918 it became clear to the War Office that with the immense growth in the importance of army communications the system under which the signals units of the army were attached to the Royal Engineers was no longer satisfactory, and it proposed to form them into a single arm. In September it sent out a memo on the subject to GHQ France. The following is an extract:

> The conditions of modern warfare are due much, if not entirely, to the revolution that has taken place in recent years in methods of transportation and communication. Though it might be said that air reconnaissance had played an equally important part,

without the requisite communications for the dissemination of such information the value of that information would be largely reduced.

This led in 1920 to the formation of the Royal Corps of Signals responsible for, among other things, army WT and RT. It might have been expected that this would mean that the battle to establish these as the first means of communication for the army had been won, or at least was well on the way to being won. However, as we shall see, that was something that was still a long way off.

To those accustomed to think of air raids in terms of World War II the Zeppelin raids of World War I may appear rather trivial. However, they did not seem so trivial to those who experienced them. There were, in fact, a total of seventy in various parts of the country and casualties amounted to 895 killed and 2,192 injured, material damage also being considerable. It followed that air defence was of vital importance and as in other fields of the war, wireless played a vital part.

The defence depended to a large extent on fighter aircraft, and in the attempt to control these by wireless a regrettable dispute arose. At a conference called by the chief of the Imperial General Staff, the Director-General of Military Aeronautics, Lieutenant-General Sir Donald Henderson, reported that he wished to experiment with the wireless control of aircraft but could not do so as the Admiralty complained that the aircraft signals would interfere with fleet communication. The objection was later withdrawn and aircraft engaged in the defence against air attacks were controlled from a station at Biggin Hill.

Like their naval forces, the German airships used their wireless both for intercommunication and for communicating with base. Weather or other conditions might change during their long, comparatively slow flights, and if this happened it might be necessary for orders to be changed. They also relied on wireless for navigation. By that time the Germans had a group of DF stations which would take the bearing of signals sent out by an airship; these would be plotted at a central HQ and the airship's position would then be transmitted to it. Unfortunately for the Luftwaffe, the defence also had DF stations and these also took bearings of the airship's signals. This kept the defence fully informed of the airship's progress across the North Sea and enabled them to estimate the probable target, so having defending aircraft ready and waiting for its arrival.

The last Zeppelin raid took place on the night of 5-6 August 1918, and wireless played an important part in the defeat of the airship force engaged. Five Zeppelins, including one which had only recently been completed, set out to attack the English Midlands. Apparently the German commanders were given incorrect information by their own wireless and thought that they had reached a point farther west than in fact they had. As a result they dropped their bombs into the sea and turned back for home before they had reached the East Anglian coast. It was fortunate for them that they did so for a large force of fighters was waiting for them. By that date the airship had shown itself no match for fighter aircraft and most would have been destroyed. As it was the L70, the pride of the German airship service, which had somehow become detached from the rest of the force, was shot down.

As the war entered its last year wireless began to be used for something new—to make peace. On 8 January 1918 the US high-power station at Newark, which had recently been equipped with a 200 kW Alexanderson alternator which gave it worldwide range, broadcast President Wilson's Fourteen Points. Then, later in the year, when it became clear that the conflict could have only one outcome and that that could not be long delayed, it was used in an attempt to short-circuit peace talks which were taking place through neutral channels. It called up the big German station at Nauen and on receiving a prompt reply transmitted a direct message to the German people appealing to them to remove the Kaiser.

In November 1918 what may be regarded as the most dramatic message ever sent by wireless telegraphy was received in London.

The end of the war was clearly imminent and in the wireless station on top of Marconi House in London a succession of operators were monitoring FL, the Eiffel Tower, Paris. As each man's watch ended he was reluctant to hand over to his relief for all wanted to be the one to receive the expected message. At 0500hr on 11 November it was received and was immediately handed to the despatch rider waiting below, to be hurried to Downing Street. It read:

From Marshal Foch to All Allied Commanders—Hostilities will cease at 11.0 o'clock.

9
Progress between the Wars

The return to peace found the position in the British Army still very unsatisfactory. Many senior officers were still suspicious of both WT and RT and were reluctant to use either when there was an alternative. In other words, neither system was regarded as a first means of communication.

The impact of radio on the conduct of the war had been immense and when hostilities ceased there was great activity in many fields aimed at putting to civilian use the great technical advances which had been made. One might have expected that, with plenty of time now available for thought and experiment, a serious effort would have been made to study the lessons learned by the use of wireless telegraphy in France to ensure that the British Army would enter any future conflict adequately supplied with equipment and with personnel properly trained to handle it.

However, there was at that time a widespread belief that there was not going to be any future conflict; the war to end war had been fought and won and there was no need to make preparations for another. We now know only too well that those who adopted this attitude were indulging in wishful thinking but many years were to pass before there was some return to reality.

The issue of CW sets to the army began during the last year of the war but it was not until 1923 that what became known as the 'A' set, a portable set with a range of six miles, designed for use in forward positions, became available. It was expected that this would meet all requirements and in the general exercises of 1924 it was decided to ban the use of line telegraphy and experiment with the use of wireless as the only means of communication. The experiment was a failure, due both to the 'A' set proving unsatisfactory and to personnel being insufficiently trained. One result of this was that the use of wireless as a means of communication in the field did not receive its much-needed boost. At about the same time there were introduced two sets designed for use inside tanks, which badly needed them. However, the range of these, two and five miles respectively, was found to be inadequate. It also proved difficult to fit them into tanks as the design of the vehicles did not allow sufficient room for them.

Whatever had been the difficulties with wireless in the trenches, there was no doubt that the artillery had found it indispensable. Nevertheless, during the postwar period the artillery had no set specially designed for it. Nor was the cavalry any better equipped, for as late as 1929 it was without a new set. The sets then in use in the army were unsuitable for cavalry, as the motor and generator would not fit on to a limber. However, the Marconi Company produced a new pack set in which the motor and generator were separate units which could be carried one on each side of a horse.

During this period the army radio had to contend with two serious difficulties. One was the growing congestion on the medium waveband, due to the development of broadcasting; the other was the difficulty of obtaining suitable sets in general manufacture. This latter led to the army having to do its own development, something in which it was severely handicapped by shortage of funds. Also, as a result of Western Front experience there was, despite the growth of motor transport, a tendency to design for transport by manpower, with consequent low power and inadequate range.

By the early 1930s the young enthusiasts of the post-Boer War period had become the Old Guard of military orthodoxy, and the struggle over wireless had to be fought again. This time it was not the cavalry that the exponents of the new-look army wanted to equip, but the tanks. Armoured fighting vehicles, to use their official designation, had made their first appearance on the Western Front in 1916, and their importance had increased as the war proceeded. At a very early stage it had been realised that a great advantage would be gained if they could be directed from the air and, as we have seen, attempts were made to develop RT communication between aircraft and tanks. Those early experiments were unsuccessful but by 1930 there had been vast technical progress and it was now quite another matter.

The Royal Tank Corps was established as a permanent arm in 1923 but at that time its vehicles did not carry radio. One of its officers, Captain B. H. Liddell-Hart, has pointed out in his *Memoirs* that ideas concerning the use of tanks based on the experience of the Western Front died slowly; what had been impossible in 1918 was still regarded in some quarters as impossible in 1935, and consequently the idea that tanks could be employed to maximum advantage if an officer in the air could communicate by RT with their commanding officer was not automatically accepted.

Up to about 1930 tanks were used mainly in small numbers to support infantry and the idea only slowly developed that they could be used independently. However, movement was hampered by the need for the officers to be continually descending from their vehicles and conferring together and as yet there was widespread opposition to the use of radio between tanks on the ground that it would be difficult to operate. Fortunately there were some in high places who did not take this view. When Brigadier P. C. S. Hobart was given command of the 1st Tank Brigade in 1932 he declared, 'If you really believe in wireless, it will work', and he gave orders for all officers in the brigade to be instructed in it. At about the same time Lieutenant-Colonel (later Lieutenant-General Sir Charles) Broad, who had agitated for the fitting of radio to tanks, worked out a simple code by which any order to tanks could be represented by two letters which could either be transmitted by morse or hoisted as a flag signal. After this, radio became standard equipment in all British tanks and Liddell-Hart, writing in 1935 about the need for cohesion in advancing formations of tanks, wrote, 'the development of wireless is a timely aid towards the reconciling of dispersion with control'.

At that time the French Army's tanks had no radio and it was not until May 1939 that the French General Staff decided that they should be fitted. This decision had been delayed so long that when war broke out very few tanks had actually been fitted.

In 1936 the War Office set up a committee under the chairmanship of Major-General Sir B. J. Jackson to review the whole question of army signals and to recommend a scale of provision of intercommunication services. The committee recommended that line telegraphy should remain the normal means of communication and that WT and RT should be provided for emergency use. The decision to give WT and RT a subsidiary instead of a complementary role had unfortunate results.

Although the cessation of hostilities in 1918 was not followed by a literal beating of swords into ploughshares, it is true that some bombing planes were converted to civilian use. One of the first uses to which these converted aircraft were put was in attempts to win the £10,000 prize offered by the London *Daily Mail* for the first direct non-stop flight across the Atlantic. The obvious starting place for such a flight was St John's, Newfoundland, the nearest point in America to the British Isles, and on 28 March 1919 the first two competitors arrived there. They were Harry G. Hawker and Commander Grieve, who brought with them a Sopwith machine fitted with transmitter and receiver. The transmitter failed on a test flight, the generator burning out as a result of the propeller fixed to it being too big; Hawker cabled to England for another generator and as persistent bad weather delayed a start it arrived in time to be fitted. Conditions did then not permit a start until 18 May. This time the generator propeller proved to be too small, with the result that the power output was too low to be of any use. The receiver was also found to be unsatisfactory, so that nothing could be received from St John's. When the attempt was at last made it soon ended in disaster, the aircraft coming down in the sea due to trouble with the engine-cooling system. The failure of the flight could not be blamed on the wireless but it does seem strange that such important equipment was not thoroughly tested before the party left England. Perhaps those concerned were still, in 1919, not yet fully wireless-minded.

A month later, on 14 June, John Alcock and Whitton Brown took off in their wireless-equipped Vickers-Vimy biplane. They had hardly cleared the land when the propeller of their generator broke off—once again it cannot have been adequately tested—and the airmen had to manage without wireless for the rest of the flight.

After this most inauspicious start to transatlantic aircraft wireless, it is pleasant to be able to record the excellent results obtained during the flight of the first transatlantic airship, R34, later that year which left Edinburgh for Lincoln, Long Island, on 2 July 1919 and arrived safely after a voyage lasting four days eleven hours. The wireless equipment comprised a low-power spark set, a CW transmitter with a range of 1,500 miles and an RT set with a range of 50 miles, and she was able to communicate with Poldhu in Cornwall, and Glace Bay, Nova Scotia.

By now wireless was closely linked with flying and it was clear that it would be called upon to play an important role in the new industry. In 1919 a commentator in *Wireless World* wrote, 'With the improvements that have now been made, the guiding of aircraft by directional wireless is rapidly becoming an accurate and reliable means of aerial navigation.' This was certainly true when in 1926 Commandant (Gen) Franco made the first crossing of the South Atlantic from Spain to South America in a Dornier machine fitted with DF equipment. The pilot relied on DF for navigation, taking bearings of the Canary Island stations en route, and after the 1,440 mile ocean crossing he made a landfall on the small island of Fernando da Noronha. Another pioneer flight in which wireless played an important part was the Amundsen-Ellsworth flight over the North Pole in the Airship *Norge* in June 1926. The equipment enabled Amundsen to obtain bearing and meteorological information while over the Pole.

The first regular passenger air service to be inaugurated was between London and Paris. This made essential the provision of a chain of ground stations for control and navigation and these were constructed early in 1920, sited at Croydon, Lympne and Le Bourget. The first flight of a civil airliner fitted with wireless took place on 4 March 1920 when Handley-Page GEALX, equipped with a Marconi transmitter powered by a 100W wind-driven generator and a five-valve receiver, left Cricklewood for Paris. Bearings and control were given from Croydon. Fog was encountered en route but Le Bourget reported all clear there.

It was not long before regulations were issued making the fitting of aircraft with wireless compulsory. In 1924 these covered all aircraft carrying ten or more persons including the crew, and aircraft flying more than a hundred miles or fifteen miles over sea were required to carry a licensed operator. As the volume of flying increased, so radio aids to aerial navigation developed. Aircraft, besides being able to fix their position by DF bearings taken from the ground, were themselves fitted with direction finders with which they could take bearings from ground stations; these could be either ordinary communication stations or special radio beacons which sent out signals at frequent intervals.

The first completely new device developed for radio navigation was the radio range. This consists of a station situated near an airport which transmits four pairs of beams: north, east, south and west. Each pair is made up of one beam which transmits the letter A (·–) and another which transmits N (–·). The arcs of the beams overlap and the two beams are transmitted simultaneously. If the aircraft hears neither A nor N but instead a long dash it will know that it is 'on the beam'; if instead it hears either A or N separately it will know that it is 'off the beam' and must change course until it hears only the long dash (Fig 7).

This system was used in reverse by both sides in World War II as a means of directing bombers over enemy territory. For this a station sited at the starting point directed a beam on the target and the bombers flew along it. Such beams could be made to give misleading information; for instance, on the occasion of one of the last German raids over Britain, when the target was apparently Bristol, the British defence was able to interfere with the German beam with the result that the raiders were completely deceived about their position and dropped most of their bombs on open country.

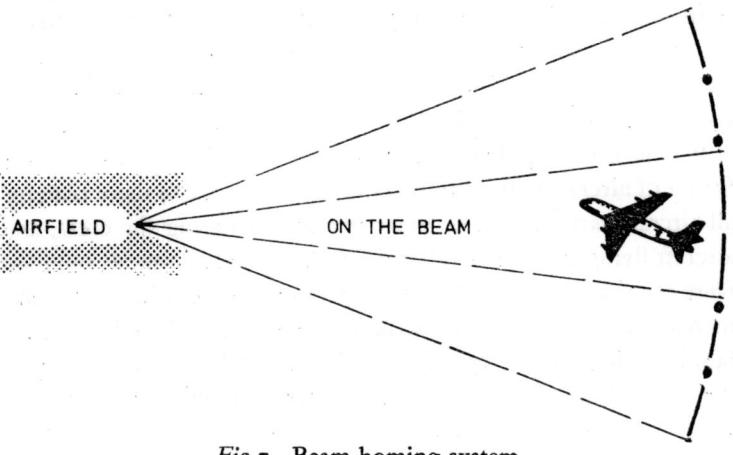

Fig 7 Beam homing system

When long-distance flying began airmen navigated by means similar to those used at sea, that is by observations of the heavenly bodies or by estimating course and speed (dead reckoning). We have seen that in 1919 Alcock and Brown had to navigate without radio and as late as 1927 Lindbergh on his first solo crossing of the Atlantic carried no radio. However, as speeds increased astronomical navigation became more difficult and consequently less accurate, so that of necessity radio aids became increasingly sophisticated and today the navigation of modern airliners depends on them completely.

Progress between the Wars

It is interesting to look at the way in which the relationship between wireless and flying has progressed. At first aircraft carried no wireless. Then military aircraft were fitted with transmitters for communicating observations to the ground. With technical improvements two-way communication became possible. Finally aircraft became entirely dependent on radio aids for their navigation. Furthermore, as the number of aircraft increased the need for ground control also grew. Today it would be quite impossible to allow aircraft to arrive and land at airports just as they pleased. Modern civil flying could not be carried on without radio and the marriage between the two technologies is complete.

Meanwhile, what was happening at sea, the area in which wireless grew up? As we have seen, the development of the thermionic valve was at first slow, and with the exception of a few Fleming diodes fitted on large liners it had by the outbreak of World War I made virtually no impact on merchant shipping wireless; nor had the position changed much by the beginning of 1919. The magnetic detectors, which when they had been first introduced had made wireless a commercial proposition and had since given some fifteen years of good service, had been replaced by balanced crystal receivers, the last disappearing just after the end of the war.

The slow adoption of the valve was at first partly due to its high cost. Initially the making of it had been a craftsman's job but the immense increase of demand for war purposes had necessitated a change to mass production methods and ample supplies were now available. From this point on they steadily replaced the crystals. The case of the transmitters was different. The prewar sets aboard those vessels which had survived the torpedoes and the mines were still capable of giving years of adequate service on ships not handling heavy traffic and too much capital was tied up in them to permit unnecessary change.

However, there were changes in new installations. With the seizing of German patents, the Marconi Company discontinued their rotary spark-gap transmitters, replacing them with quenched gaps, hitherto the main feature of the Telefunken system, and they introduced two new ship transmitters which were both less bulky and more efficient than the old ones. For the passenger ships they introduced a CW transmitter which made it possible for ships on the North Atlantic run to maintain contact with shore stations throughout their voyages. Later came short-wave valve transmitters which had, under favourable conditions, a worldwide range. These went not only into passenger ships but also into tankers and the more important freighters, whose owners found it very advantageous to be able to communicate with them quickly, wherever they might be.

Of great convenience to passengers was the development of ship–shore radio telephony. In 1920 an experimental installation was fitted aboard the *Victoria* and during her voyage to Canada highly satisfactory tests were carried out with the Marconi stations at Chelmsford and Poldhu. With the ship about nine hundred miles from Chelmsford a concert was received and passengers in the first-class lounge were able to hear it from a loudspeaker, the first time that this had been possible aboard a British ship. The fitting of RT to passenger ships began in 1922 and the service was soon extended to other vessels.

Another important development about this time was the fitting of ships' lifeboats with wireless telegraphy. A start was made with this in 1914, when some of the lifeboats carried by the *Aquitania* were fitted. After that several other large liners had their boats similarly equipped though there was no general move in that direction. Then in 1923, nine years after the practicability of fitting lifeboats had been demonstrated, there was the affair of the *Trevessa*.

The *Trevessa* was a small tramp belonging to Messrs Hain's of Cardiff. While on a voyage from Australia to the UK loaded with zinc concentrates—an extremely heavy cargo—she sprang a leak in a very lonely part of the Indian Ocean. It soon became apparent that she was going to sink and the captain ordered the wireless officer to send out an SOS. After some delay a reply was heard from the *Runic*, whose signals however were very weak, indicating that she was a long way away. It was impossible for the crew to remain on board until help from her arrived, and with her end imminent the order was given to abandon ship. The two lifeboats were lowered without mishap and the *Trevessa* sank.

It was clear that it would be useless for the boats to attempt to remain where they were until the *Runic* arrived, as inevitably they would drift, so an attempt had to be made to reach land. The captain's boat reached Mauritius, 1,556 miles away, after twenty-two days, and the chief officer arrived at Rodriguez Island, 1,747 miles, after twenty-seven days, which was then a record for an unbroken passage in an open boat. If either boat had been fitted with wireless it is reasonable to assume that it would have been possible to have established communication with the *Runic* or some other ship, and the two lifeboats would have been spared those long and hazardous voyages, which might well have ended with the loss of a great many more lives than the one that actually was lost. This incident created a great impression in shipping circles and the Board of Trade issued regulations requiring certain passenger vessels to carry at least one lifeboat fitted with wireless telegraphy; by 1926 the Marconi Company had fitted 161 lifeboats.

It has been said that radio makes every ship equipped with it a potential lifeboat but in order for this to become literally true it is necessary for every ship radio station to be continuously manned or, where this is not economically practicable, for it to be equipped with some means of calling a single radio officer when he is off watch, for disasters do not occur only at scheduled times. The problem which had to be solved here was in fact a very complex one. To begin with, the receiving instrument had to fulfil two requirements: first, it must respond to every distress call that would normally be in range; second, it must respond to nothing else, for too many false alarms would have the same effect as crying 'wolf' too often. Furthermore, it must function satisfactorily in both respects in busy areas where there is an almost continuous background of signals at all strengths coming from many stations, and also under typical conditions where static much heavier than is heard in temperate zones goes on all and every night. From either of these backgrounds or from a mixture of both the distress signal must be extracted. A further complication is that attempts to pick up SOS are confused by the fact that fortuitous combinations of the three letters are by no means uncommon. They occur with such combinations as 'so slow' and many others, not to mention a great many combinations of letters such as IJS, SMB or EUGI, which are made up of the same series of dots and dashes and which may be run together with bad spacing.

For years engineers wrestled with the problem and several instruments were submitted for official approval, but without success. One of these transmitted a series of dots at the rate of three a second, regulated by a flywheel. For another it was claimed that it would respond to SOS and/or any other desired signal. Neither of these, nor others, received official approval. Then in 1920 the Marconi Company demonstrated a calling device or 'auto-alarm' which would respond to a series of three four-second dashes, causing alarm bells to ring. This was exhaustively tested and finally given official approval. But before the auto-alarm could be brought into use it was necessary to alter the regulations regarding distress calls and a new signal, known as the alarm signal, was introduced. This consisted of a series of twelve four-second dashes with intervals of one second, it being considered that the instrument would respond to at least one series of three out of the twelve. This had to be transmitted before the SOS, a short interval being given between the two in order to allow the operator to reach the receiver. Fitting of this instrument to all vessels carrying fifty or more persons became compulsory in 1927.

The auto-alarm was extremely complicated and delicate and space does not permit a description of it here. When first fitted it did not always fulfil the requirement not to give false alarms, and sometimes, especially in tropical waters where static was heavy and continuous, its three powerful electric bells were heard much too often. But in due course its teething troubles were overcome and it became a standard piece of equipment.

A most important addition to the services rendered to the mariner by wireless telegraphy was the introduction of the ships' medical service. Only very few freighters carry a doctor, responsibility for giving medical care to any member of the crew who may need it resting with the captain, who usually delegates it to a deck officer or the chief steward. Every ship must carry a medicine chest containing drugs and instruments as officially listed. There must also be a copy of the *Shipmaster's Medical Guide*, an excellent manual giving instructions for the diagnosis and treatment of the simpler ailments and injuries likely to be encountered at sea. Deck officers also have to pass an examination in medical knowledge.

The standard of health at sea is high; nevertheless there are inevitably cases for which the above facilities, excellent though they are, are insufficient. An arrangement was therefore arrived at by which a captain requiring medical advice can send a message describing the patient's symptoms to a ship carrying a doctor. The doctor is able to prescribe treatment, bearing in mind the contents of the standard medicine chest. In serious cases arrangements can be made by wireless for the two ships to meet and for the sick man to be transferred to the ship with the doctor. In the 1920s, it must be remembered, there were no helicopters to pick up a sick man and take him to hospital.

As we have seen, during World War I tremendous advances were made in the three-electrode valve and in the means of manufacturing it. The years that followed saw the development of a vast technology based upon it as it superseded the spark system of wireless telegraphy. Nor was the spark transmitter the only one to go. Valves capable of handling considerable power were developed and these were built into transmitters which displaced the timed spark transmitters, the big alternators and Poulson arcs, on which long-distance wireless had come to be based.

Parallel with this revolution was the growth of the use of short waves. It is often said that the potentialities of the short waves were first discovered by the amateur experimenters, who had been crowded off the medium waves and given the short waves which nobody wanted. It is perfectly true that they were given short-wave bands, and that they did indeed make very good use of them, achieving some spectacular results, but they were not the first to use them. The fact that waves could be reflected was known from the beginning; indeed it had been discovered by Hertz, who used parabolic mirrors to reflect waves of one metre and less. In 1896 Marconi had used reflectors when demonstrating his system from the top of the Post Office in London, and in 1897 he had placed metal sheets on each side of his transmitter to act as reflectors during his experiments on Salisbury Plain. However, after this not much was heard of reflecting waves for some time. This was because there is a definite relationship between the length of the wave used and the size of the reflector needed to reflect it; for medium and long waves the reflectors would have to be too large to be practical and it is therefore only possible to use them for short waves.

Marconi also used short waves in his early experiments but a belief grew up that for long-distance communication long waves were more efficient, and for the very high-power stations wavelengths from 8,000m up to nearly 20,000m were used. However, in 1916 Marconi turned his attention to short waves, resuming experiments with them and with the use of reflectors. His results were most encouraging. He continued experimenting in this field until nearly the end of his life, and in 1935, working in Italy and transmitting with a wave of only 35cm, he obtained reflections from a car moving along a road about a mile away—in fact he had entered the field of radar.

Progress between the Wars 143

As soon as the practicability of the beam station had been established it was put to commercial use; the first service employing it was opened in 1922 between Ongar in England and Berne, Switzerland.

We must now return to the imperial chain. We saw that the original scheme for the establishment of a chain of high-power wireless stations linking the far-flung parts of the British Empire never really got off the ground, due first to political wrangling and then to the outbreak of World War I. But with the cessation of hostilities it became clear that the need for such a chain still existed. Interest in it revived and with it controversy. However, there had been revolutionary technological developments since the original designs had been made and these were now obsolete.

By 1920 large valve transmitters had been developed and the range of high-power stations had been considerably increased. Nevertheless, in that year the Imperial Wireless Committee reported that the stations should be 2,000 miles apart and that they should be built by the Post Office construction department. The first station of this chain was built at Leafield near Oxford, and opened on 18 July 1921. But technology was getting ahead of political decisions and the only other station to be built under the scheme was at Cairo. Controversy continued, without any improvement in imperial communications. Britain was falling behind in this field so that by the middle of 1923 the position with regard to high-power, government-owned stations was as shown in Table 5.

TABLE 5

Country	*High-power stations*
United States	10
France	10
Great Britain	2

Moreover, the two British stations, Leafield and Cairo, were by then obsolescent.

The opening of the beam service to Switzerland was followed by a development on a much larger scale. In July 1924 the Marconi Company entered into a contract with the Post Office to erect beam stations for direct commercial services to Canada, Australia, India and South Africa. When these were opened wireless telegraphy may be said to have entered into full competition with the cables on a worldwide scale. And it was a very serious form of competition; the cost of erecting and maintaining a pair of beam stations was much less than that of several thousand miles of submarine cable. The situation which the Anglo-American Cable Co had feared a quarter of a century earlier had come about. The Canadian beam service was opened in October 1926, the others during the following year, and by the end of 1927 in competing areas the cable companies had lost half their business and were in serious difficulties. The British government became alarmed as for strategic reasons it could not let the cable system break up, and so stepped in. As a result a merger between the wireless and cable interests was negotiated and a new company, Cable & Wireless Ltd, was formed to take over Marconi's Wireless Telegraph Co, The Eastern Telegraph Co, The Western Telegraph Co, and the British Post Office's cable and beam services. Considering that the Marconi Company was in by far the stronger position it is not at all clear why it accepted the position of junior partner. After World War II Cable & Wireless Ltd was nationalised by the Labour government.

These stations were not erected as part of the imperial chain but to a large extent they achieved its aims, one in particular proving of immense value during World War II. Many of the cables in the Mediterranean were cut during that conflict, seriously impeding communication between London and the desert army in North Africa. However, wireless was able to fill the gap and nearly the whole of the traffic with the desert army was handled by the beam station at Wincanton in England.

Meanwhile, in 1923, one more government committee had reported on the imperial chain and as a result the Post Office proceeded with the erection of the giant station GBR at Hilmorton, Rugby. This station comprised transmitters for telegraphy and telephony and its twelve 820ft masts were soon familiar to travellers on the main London to Crewe line. The telegraphy transmitter worked on a wavelength of 18,740m (16kHz). It had worldwide range and when it was opened the total power used by both telegraph and telephone transmitters was 1,400kW.

Apart from being able to communicate with fixed stations abroad, this station gave a valuable service to shipping. It at last provided a British time signal and it also broadcast a news service. Most important, though, was its long-distance message service; this enabled a sender to hand in a telegram addressed to a ship whose whereabouts were unknown. The message would be broadcast twice from Rugby and as all British ships listened to Rugby's twice-daily schedules, it would be received wherever the ship might be.

Most of the developments with which we have so far dealt originated in areas of high population density, in Europe or America. We must now say something of one that had its birthplace in a country of wide open spaces, Australia. It was here that radio telephony, as distinct from broadcasting, made its greatest social impact. To the people of the outback living on sheep stations and in shacks, often long distances from any neighbours, it was a gift from heaven.

However, before it could be fully made use of a serious difficulty had to be overcome: many if not most of the isolated homesteads had no electric power. To bring batteries from a distant township was often not practicable and some new source of power had to be found; the answer to the problem was pedal wireless. This ingenious but simple device was developed in Adelaide in 1929 and consisted of a small generator and a construction of bicycle parts which enabled the user to produce current by pedalling. This machine enabled the people of the isolated homesteads to summon help from distant neighbours in an emergency, or to call

up the flying doctor for advice or a visit; it made it possible for
the children of the homesteads to be given lessons by wireless;
perhaps more than anything else it affected the women of these
lonely places. Before pedal wireless they might go for long periods
without speaking to anyone of their own sex; now they could
call up a neighbour when they felt lonely and have an over-the-
garden-fence chat like any suburban housewife. In fact it was
possible for several to join in the conversation, the only thing
missing being the coffee or tea. Not everyone was happy about it,
however; scornful old-timers declared that wireless had reduced
the outback to the triviality of the suburbs!

This period saw the beginning of the use of radio in yet another
field, one in which it has become of great importance. T. A.
Crichley, in his *A History of the Police in England and Wales, 1900–
1966*, describes the introduction of radio into the police service.

As we have seen, wireless telegraphy was first used in the
detection of crime in 1910, but this only involved the use of the
ordinary public wireless service. The Detroit Police were the first,
in 1921, to experiment with the use of WT for intercommunication,
while in Britain operational use started in 1923 with the fitting of
a van with WT by the Metropolitan Police. This, however, does
not seem to have been very successful, though when the General
Strike occurred in 1926 the Lancashire Police found a van fitted
with WT very useful. By this time several forces were carrying out
experiments.

It is the Brighton Police Force which deserves the credit for
the first 'walkie-talkie' receiver that could be carried by a con-
stable. This was first tested in 1928 and it was introduced into
the police service in 1932, when it became known as the Brighton
Receiver. A big step forward was taken in 1934 by the Metropolitan Police, when they set up an information-room which was linked
by WT to fifty patrol cars; but progress was still slower than could
have been desired, due mainly to the fact that only the medium
waveband could be used, VHF channels being not yet available.

10
Words and Music: Part One

So far we have concerned ourselves almost exclusively with wireless telegraphy (WT), which we may define as the wireless transmission of signals, though we have said a little about the beginnings of radio telephony (RT), the wireless transmission of speech and music, and about broadcasting, transmission to a multiple audience. We must now trace the development of RT and broadcasting in some detail.

From the very first the experimenters in the field of RT did not distinguish between the transmission of speech and of music, and thought in terms of reception by a multiple audience. It follows that the history of wireless telephony is inextricably bound up with the history of broadcasting. Actually the idea of broadcasting speech and music pre-dates the development by Marconi of the first practical wireless telegraph, going back to the appearance of the first telephone. That instrument was invented by Alexander Graham Bell, the son of an Edinburgh educationalist. Bell was appointed professor of vocal physiology in Boston University, which involved teaching deaf mutes to speak. There he studied the transmission of sound by electricity and in 1876 he produced his telephone.

Bell had no sooner secured a patent for his invention than he turned his attention to the transmission of music and song. In 1879 he transmitted music, including operatic arias, from Boston to Providence, Rhode Island. Later he transmitted a programme which included an address by the evangelist Dwight L. Moody and a hymn accompanied by I. D. Sankey, which may be regarded as the absolute beginning of religious broadcasting. Next, in 1890, guests at a party held in the home of a telephone official danced to music received by telephone. It might have been thought that these developments would have gone ahead more quickly than they did, but the Bell Telephone Co, which had taken over the exploitation of Bell's invention, were not interested in them. They regarded it as much more profitable to expand the sales of telephones for two-day conversations.

Interest in what may be called line broadcasting was not confined to America. In 1881 a Hungarian, Theodore Puskas, demonstrated the possibility of transmitting speech and music to a number of listeners simultaneously, and he established a 'telephone newspaper' in Budapest. Interest was also aroused in Great Britain, and in 1894 the Electrophone Co was formed in London. This was conspicuously advertised in the telephone directories of the years during which it existed. Subscribers were supplied with four sets of headphones and an 'answering-back microphone', and could call the Central Office and ask to be connected to any of the theatres, churches and lecture halls fitted with microphones. It is said that reception was often not good, which is hardly surprising when one considers what microphones and other equipment were then available, and the venture was not a commercial success. One cannot help feeling, though, that with their answering-back microphones the subscribers had something denied to present-day listeners.

Interest in line broadcasting petered out, to be revived years later in the form of radio relay, an adjunct to radio.

During the period prior to World War I advances in technology in Britain and Europe were mainly concerned with wireless telegraphy. The early development of radio telephony and broadcasting took place largely in the United States, where it depended principally on the work of the small group of inventors we have already mentioned—Fessenden, De Forest, Poulsen, Alexanderson and a little later, Armstrong; like the group which launched wireless telegraphy, a cosmopolitan band. Nevertheless, Sir Oliver Lodge in Britain seems to have been one of the first to see that a wireless transmission could be directed to a multiple audience. Addressing the select committee appointed by the House of Commons to consider the report of the International Radio Conference in 1904, Lodge said, 'It might be advantageous to shout the message broadcast to receivers in all directions, such as for army manoeuvres or sporting events.'

We have already noted Fessenden's historic broadcast on Christmas Eve 1906. De Forest also devoted himself to broadcasting, and in 1908 he and his wife crossed to Paris and obtained permission to broadcast from the Eiffel Tower, giving a programme of records which was heard over a considerable area.

Public interest in radio continued to grow and throughout the United States a considerable number of people, mostly young, became enthusiastic experimenters in both transmitting and receiving. Some, outstandingly successful, put on regular transmissions every week or so, with programmes including news and music, though there was still one serious difficulty, the lack of a satisfactory microphone, the only type available being the carbon microphone as used in ordinary telephony; the carbon granules it contained heated up and as the transmission proceeded they stuck together, with the result that quality steadily deteriorated and a transmission could not last much more than an hour.

The value of broadcasting as a means of spreading ideas seems to have been realised at an early stage, and it is perhaps not surprising that it was a member of De Forest's own family who first took advantage of this facility. In 1909 his mother-in-law, Mrs Harriet Blatch, spoke 'on the air', making women's suffrage the topic of the first political broadcast.

An important milestone in the progress of broadcasting was reached when De Forest erected a transmitter on the roof of the Metropolitan Opera House, New York, and on 12 January 1910 broadcast direct from the stage a concert given by Enrico Caruso. It was listened to by a considerable number of people, many sharing headphones, and it may be regarded as the beginning of outside broadcasting. However, as yet the growing radio industry showed little or no interest in anything that we would now call broadcasting, though there was one young man in the service of a major company who could see farther than his superiors: this was David Sarnoff, whom we have already mentioned in connection with the *Titanic*. In 1916, an engineer advancing rapidly in the service of Marconi's Wireless Telegraph Co of America, he saw the potential of radio for bringing entertainment into every home and suggested the formation of a company to found what would be a new industry, providing programmes and manufacturing receivers for the home. But nobody was interested. There was more money in supplying the allies with military radio equipment; then the war reached America and Sarnoff had to shelve the idea.

With the ending of World War I all forms of radio communication entered upon a period of great change and 1919 was to see the dawn of a new era with the three-electrode valve completely changing its technology. Serious experimental work had begun to move from the garret and garage stage to the laboratories of large commercial undertakings and new developments were now no longer linked with the names of individuals but with the big corporations whose staffs had been responsible for them. And very soon wireless, or radio as we shall now call it, began to make contact with the non-technical mass of the population. Wireless or radio became 'the wireless' or 'the radio'; in other words, broadcasting was coming into existence.

However, before these changes began to have much impact on the general public, a major reorganisation of the radio industry took place. It was felt that American radio was too much under the control of foreign (that is to say, British) interests; the Marconi Company of America was British-based, and with its ownership of many basic patents was in a very strong position. The situation was not satisfactory. Negotiations took place and as a result the Radio Corporation of America (RCA) was formed to acquire the radio interests of Marconi's Wireless Telegraph Co of America, the General Electric Co, the Western Electric Co and the American Telephone & Telegraph Co. The only important American company left out was Westinghouse which became associated with RCA later.

The history of broadcasting is even more complicated than the history of wireless telegraphy or of straight wireless telephony and we can only record briefly the more important events and developments and note their impact on the community at large. For those who want to study the subject more thoroughly there are two admirable works, *The History of Broadcasting in the United Kingdom* by Asa Briggs, and *A History of Broadcasting in the United States* by Erik Barnouw. Here we shall begin with the development of broadcasting in the United States, which was, in fact, its birthplace.

From its inception wireless telegraphy enabled a person at one point to communicate directly not only with one person within range of the equipment but also simultaneously with any number of suitably equipped persons within range. In that sense 'general calls' were from the start a form of broadcasting, but it was necessary for the receiving points to be manned by trained staff. Radio telephony did away with this limitation, as its receiving instruments could be used by anyone. With its invention radio ceased to be something for professionals only and became open to the whole community. Another development was that during the war the big American electrical companies, such as General Electric, Western Electric and Westinghouse, had erected large plants to turn out vast numbers of transmitters and receivers, while the electric lamp manufacturers had installed plant for producing valves (vacuum tubes) on a large scale. This capacity was now idle.

Soon after the freeing of radio from wartime restrictions, the

Detroit News acquired a radio station from which it broadcast news and music to the surrounding district. When, early in 1920, the presidential primary elections took place, it broadcast the results as they came in. However, the power used was low and the number of people who could receive the broadcasts was limited. There were in fact a number of such stations about the country broadcasting intermittently on low power. One was at Pittsburgh, owned by Frank Conrad, an engineer employed by Westinghouse, who broadcast frequently from a transmitter in his garage. On one occasion Conrad included in his programme an advertisement for a particular item in a Pittsburgh department store and it immediately sold out. This was one of the earliest demonstrations of the value of radio as an advertising medium.

Erik Bernouw has described the events leading up to the launching of commercial broadcasting in the United States. Early in 1920 Conrad's broadcasts came to the attention of one of Westinghouse's vice-presidents, Harry P. Davis. The company, like others in the industry, was looking for new fields in which to employ its idle capacity. Davis noted that the broadcasts were being received only by people who had themselves constructed their receivers; there must, he thought, be many people who would like to hear them but who were unable to build a receiver. Why should not Westinghouse build receivers to sell to them? Here was the field they were looking for. There was one stumbling block. The broadcasts were being made at irregular intervals while what was needed was a daily service with the time of transmission advertised. He at once set about launching the new product, a receiver to be marketed under the name of 'the Music Box'. The presidential election would be taking place in November and then, surely, would be the time to inaugurate the new service. Davis instructed Conrad to build a more powerful transmitter, which must be ready for the election, and he also applied for a licence to transmit. This was granted, together with the call sign KDKA. Finally, arrangements were made with the *Pittsburgh Post* to supply the election results as they came in.

At 2000hr on 2 November 1920 the first American commercial broadcasting station, Pittsburgh KDKA, went on the air. Its power was 100W and it was estimated that there were between 5,000 and 10,000 receivers able to receive it. From that date the broadcasts kept to a regular daily schedule, at first one hour daily, then, as interest grew and power was substantially increased, more and more. The cost of running the station was, of course, born by Westinghouse, as Davis saw that the company would be amply reimbursed through the sale of receivers and from the advertising value of keeping its name constantly before the public.

Several of the items which have traditionally made up radio programmes had their 'firsts' during the few months following the opening of KDKA. On 2 January 1921 the station put out the first religious broadcast when it relayed a service from the Pittsburgh Calvary Baptist Church, a programme which may also be regarded as the first commercial outside broadcast (OB). The first political broadcast from a commercial station took place on 15 January of the same year when Secretary of Commerce (later President) Hoover spoke from KDKA, while the first sports OB took place in April when a boxing commentary was broadcast. In May the first relay from a theatre went out. However, the most spectacular of these 'firsts' was not from KDKA. David Sarnoff, who had recently been appointed general manager of RCA, set up a temporary station in Jersey City and on 2 July 1921 broadcast the first running commentary of a big fight, that between Jack Dempsey and Georges Carpentier. For the fight Tex Rickard organised a hundred centres where receivers with loudspeakers were set up; a charge for admission was made and the proceeds sent to charity.

During 1921 interest in radio spread like a prairie fire over the whole of the United States. Between October and December 21 new stations were licensed; in January to March 1922, 67; April to May, 187. By March 1921 there were 50,000 receivers in the United States; by May 1922 there were 750,000. This proliferation of stations was not checked by any official restraint, for under the 1912 Radio Act the secretary for commerce had no powers to refuse a licence to any American citizen. Compared with the hesitant and restrained launching of broadcasting in Great Britain, the early radio scene in the United States was a free-for-all and by the end of 1924 no fewer than 1,400 broadcasting stations had been licensed.

Who applied for licences and acquired stations? There were several distinct groups. One of the first in the field was that composed of educational establishments, mainly universities, which saw in the new medium a way of spreading education. Department stores were also among the early starters, using them for advertising purposes, self-advertising that is to say, not selling time on the air to other advertisers. Self-advertising was also generally the motive for the numerous newspapers which entered the field. The strangest group was that formed by what became known as the 'ego' stations, that is stations bought by rich individuals for their own use. It is interesting to speculate on what the owners of these stations thought they were doing. Were they merely expressing their personality to a wider circle? Did they feel an urge to advance some great campaign for the betterment of mankind? Or did they simply want to show people what great guys they were?

The power of these stations was by modern standards usually very small, mostly 10–500W. This was just as well, for they had to share only two wavelengths—360m (833·3kHz) for news, entertainment, etc, and 485m (618·3kHz) for meteorological reports and other official announcements. It frequently happened that there were two, sometimes several, stations covering the same area and so it was necessary for them to enter into voluntary agreements to share broadcasting time. The result was that many licensees found it impossible to broadcast without excessive interference while others found it difficult to obtain sufficient programme material and to meet the running costs of their stations. Consequently, many of the stations were very short-lived, some in fact lasting only a few days, which resulted in a considerable disparity between the figures given for the number of licences granted and the number of stations actually in operation.

Not only did broadcasting arouse immense public interest as the latest marvel of the age; it was also soon appointed Public Whipping Boy Number One. If the weather was unseasonable, it was due to the radio; if crops failed, it was the radio; if cattle died for no apparent reason—well, what else could it be? Later, of course, such disasters would be attributed to the H-bomb or to a landing on the moon.

At first programmes consisted mainly of records and news, together with a few special items. Plays, when broadcast, were relayed straight from a theatre. It soon became clear, however, that radio required a technique of presentation specially tailored to its needs; for instance, silences must be avoided, nor could programmes be introduced in a haphazard way if they were to attract the maximum of interest. It was seen too that a play suitable for the stage did not necessarily suit the microphone. Soon, schools for announcers were being set up and radio playwriting competitions organised. The artistes and other broadcasters providing the programmes were at first unpaid. There were plenty of amateurs who were happy to do it for the fun of the thing while professionals were glad of the publicity. However, as demand expanded, the time soon came when the supply of such talent began to dry up; frequently too unpaid artistes failed to turn up for the programmes for which they were scheduled, and then, as a first step, the Musicians Federation demanded payment for its members' services.

Meanwhile the chaos on the two permitted wavelengths was growing steadily worse and by the beginning of 1923 it was barely tolerable. Clearly something had to be done, and at one time there were as many as twenty measures designed to regulate broadcasting awaiting attention by Congress. However, the situation was too urgent to await the outcome of the discussion of new legislation; Secretary Hoover, the minister responsible, found himself under very strong pressure to do something at once and he responded by allocating, as from 23 May 1923, the wavelength band from 222m (135·1kHz) to 545m (550kHz) to broadcasting.

This new arrangement did much to reduce the chaos on the air but it did nothing to ease the financial difficulties in which many station-owners were beginning to find themselves as more and more broadcasters began to demand money for their services. In many cases there was just no more money to pay the broadcasters and it became clear that no satisfactory solution to the problem could be found until an answer was found to the related question: in the last analysis who was going to pay for broadcasting and how should the money be collected? There was considerable support for the adoption of what had already become known as the British system, payment by licence, while another suggestion was that a tax or royalty should be added to the price of receivers, on the basis of so much per valve holder. At the same time there was strong support for the claims of advertising, as there was never the opposition to it that there has always been in Britain; nevertheless, this suggestion was by no means universally popular, even in commercial circles. David Sarnoff of RCA was strongly opposed to it, and in the political sphere Secretary Hoover declared in 1922: 'The ether is a public medium and must be used for the public good. It is quite inconceivable that it shall be used for advertising', a declaration that might very well be included in the list of 'famous last words'.

In the minds of most people in Great Britain broadcasting and advertising in the United States are closely linked. The truth is that the situation which exists today, in which most American broadcasting is financed by advertising, took several years to develop. As we have seen, the early stations were mainly owned by individuals or bodies which had some particular reason for wanting to be heard and in so far as broadcasts did in fact amount to advertising, it was self-advertising. As a general rule paid outside advertisements were not accepted. Furthermore, and rather surprisingly, advertisers as a body took some time to appreciate the value of radio as a medium; there was no pressure group of advertising interests, as has existed in Britain since the early days of broadcasting, urging the adoption of advertising in broadcasting, although it was put forward in some quarters as a means of solving the increasingly difficult problem of financing broadcasting.

It was really the American Telephone & Telegraph Co which introduced the system by which an advertiser could 'buy time on the air'. Its subsidiary, Western Electric, had a large number of orders for broadcasting stations to be erected in the New York area and it was known that the people who wanted to erect these stations were all in one sense or another advertisers. At best each would only be able to use his station for a short period each week; at the same time there were many advertisers who did not want to become involved in running a station. So why not have just one station and let out time on it to anyone who wanted to broadcast, in exchange for a toll, such as was made for a telephone line? The company started with its station New York WEAF, which opened on 3 August 1922, but there was no rush for its facilities and it was not until 28 August that the first paid commercial went out. Very little business was done during the rest of the year but after Christmas it 'caught on' and the demand for time on the air became brisk.

The first simultaneous broadcast by two or more stations linked by a telephone line took place in 1923 when a station in New York was linked with one in Chicago to enable the commentary on a football match to be heard in both cities. During the next year or two more of the technical problems encountered were solved, and groups of two or three stations began to grow into large networks, covering the whole country. The first of the giants, the National Broadcasting Co (NBC) was formed in 1926, owned jointly by the General Electric Co, Westinghouse, and the Radio Corporation of America, with the American Telephone & Telegraph Co supplying the linking lines. It was soon to have a competitor, the Columbia Broadcasting System (CBS), which grew out of the concern of a Philadelphia concert agent at the threat of monopoly by the NBC consortium. Its first programme went on the air in September 1927. At first its financial resources were tiny compared with those of its giant competitor but after a rough period it succeeded in establishing itself on a firm basis. Both networks were organised on the basis of selling time to sponsors which enabled them to pay artistes adequately and, as they developed, to finance expensive programmes.

A few figures will illustrate the tremendous growth of the radio industry during this period. By 1924 the sale of radio sets and parts had reached the sum of $358 million. Four years later it had reached $650 million. At the same time congestion on the air continued to grow until a point was reached where Chicago alone had forty stations. Again something had to be done, and Secretary Hoover announced that all possible channels had now been filled and no new stations would be licensed. Immediately all existing licences acquired a cash value and commercial concerns wishing to open stations started buying up small non-profit making stations, such as those operated by the churches and universities.

Broadcasting in America was never plagued by any ban on controversy which so hampered the BBC during its early years and it was therefore able very quickly to make itself a powerful force in politics. Like the entertainers, the politicians were quick to realise the value of the publicity offered by radio. President Harding was the first head of state to make use of the new link-up facilities to address a large audience. On 21 June 1923 he spoke in St Louis on the proposed World Court and his speech was broadcast from four stations in St Louis, New York and Washington; but he was not a good broadcaster. His successor, Calvin Coolidge, was more successful; he had a good microphone voice and he used it to address audiences of a size hitherto undreamed of.

By the 1930s radio had become a vital force in politics, with the parties spending huge sums on buying time on the air in order to put over their policies and the personalities of their candidates. Nobody understood the value of radio better than President Franklin D. Roosevelt, who on 12 March 1933 gave his first fireside talk in which he spoke about the banking moratorium. Public reaction demonstrated to him the power of radio as a means of communicating with the people and in his first year of office listeners heard him twenty times. He was in fact such an excellent broadcaster that he became known as the Radio President.

This freedom in the broadcasting of controversial matters did not mean that there were no attempts at censorship, nor that those responsible for programmes did not have some difficult moments. One of these was when a woman broadcaster managed to insert into her script a plea for birth control, a decidedly controversial subject at the time; and the organisers lived in constant fear that some 'red' should be accidentally given an opportunity to broadcast. The matter of censorship came to the fore in 1927 during the passage of the Radio Bill. A Chicago station had broadcast part of the trial, at Dayton, Tennessee, of a schoolmaster for teaching evolution, and an attempt was made to insert a clause in the bill prohibiting broadcasts on this thorny subject. The attempt failed, and when the act became law it laid down that 'the licensing authority shall make no regulations which interfere with the right of free speech', and that 'stations must give equal facilities to all parties to broadcast'.

In 1927 also there came a development which the pioneers of radio can hardly have foreseen, when a start was made with the fitting of receivers into automobiles. This new development soon grew into a considerable business, so much so that the giant General Motors Corporation launched a subsidiary, General Motors Radio, to deal with it.

Broadcasting from overseas to the United States began in 1930, the first occasion being the five-power conference being held in London. The highlight of these broadcasts, as far as listeners to CBS which arranged them were concerned, was the opportunity it gave them to hear live the voice of King George V when he opened the proceedings. Unfortunately at that time CBS had a rule that there must be only live broadcasts, with no recordings, and comparatively few people heard the broadcasts, because of the five-hour time difference between London and New York which put them into an off-peak period.

One of the most remarkable of these early foreign broadcasts took place in 1934. Many middle-aged people will remember the BBC's broadcasting, on a number of occasions, of the song of a nightingale from a wood in Kent but it is safe to say that few of them knew that one of these broadcasts was relayed across the Atlantic to listeners all over the United States. This broadcast was voted in America the most outstanding of the year. Another interesting long-distance relay in the same year was from Little America, the American station in the Antarctic. This was picked up by Buenos Aires, relayed to New York, and then again to London, a total distance of 9,000 miles.

Alexander Kendrick, in his *Prime Time, the Life of Edward Murrow*, describes how during the years of depression in the early 1930s broadcasting attained a peak impact on the general public in the United States. Eight million people were out of work; with nothing to do, nowhere to go, and often weighed down by despair, they would gather round their radios for free news, free entertainment and free comfort. In those years radio fulfilled a need which nothing else could fill. And when better times returned it retained its influence, a fact shown by the growth in the number of receivers in use:

    1927   6·5 million (equal to 1 set to 17 persons)
    1940   50 million (equal to 1 set to 2·6 persons)

The lean years had made radio a necessity of life.

11

Words and Music: Part Two

The story of broadcasting in Great Britain and other European countries does not go back so far as it does in the United States, largely because it was in that country that most of the early development of CW transmission took place. However, not all the development was American. C. S. Franklin of the Marconi Company, and A. Meisber in Germany, discovered independently that the De Forest three-electrode valve could be used to produce CW oscillations and so make practicable commercial low-power radio telephony. Marconi entered this field just before the outbreak of World War I; in 1914 he carried out RT experiments aboard the Italian warship *Carlo Alberto*, achieving a range of forty-four miles. Early in the same year experimental transmissions of speech were made from Marconi House, London, and the Marconi Company started manufacturing receivers for RT.

In Britain as in the United States the war ended with radio technology having reached a point at which all was ready for the launching of broadcasting on a commercial scale. But in Britain opposition from vested interests and organised resistance to change was arrayed much more formidably than on the other side of the Atlantic. The British—and other—governments were reluctant to give permission for broadcasting, on the ground that it would interfere with service and other official transmissions. And it must be admitted that having regard to the lack of selectivity of receivers at that date, this might have happened in some areas though not to an extent that would justify the objection. However, as in the United States, there were a number of large manufacturers who found themselves with plant set up for producing radio equipment for war purposes, but for which they now had no use, and they were able to exert pressure on the government to allow broadcasting.

The first practical steps towards the establishment of a broadcasting service in Britain were taken when the Marconi Company set up an experimental RT station at Ballybunion in County Kerry, Ireland, from which in March 1919 it carried out successful RT experiments with Louisburg, Nova Scotia, using a power of 2½kW and a wavelength of 3,800m (789kHz). Following this, a 6kW transmitter was built at the company's Chelmsford works (see p 104). Test transmissions of voice and speech were made from this, and when they proved successful power was increased to 15kW, the wavelength being 2,500m (120kHz).

The Marconi Company was granted a temporary licence by the Post Office and, starting from 23 February 1920 and using its Chelmsford station, it broadcast two half-hour programmes of speech, music and news regularly every day. These broadcasts continued for a fortnight, the artistes taking part being mainly members of the station staff, and pre-dating as they do the opening of Pittsburg KDKA they can be claimed to be the first regular daily broadcast service in the world. It was one of these programmes that was picked up by the liner *Victorian* and relayed to passengers, as mentioned in Chapter 9.

These broadcasts attracted a great deal of interest, especially from the *Daily Mail*. That paper not only began to feature broadcasting news but it organised a special concert to demonstrate what could be done. Lord Northcliffe, then the owner of the paper, was not one to do things by halves and he decided that Dame Nellie Melba, then one of the most prestigious figures in opera, should be engaged to sing. The concert was broadcast from Chelmsford on 20 June 1920, with an exceptionally distinguished studio audience. The receiving audience was of course limited to those already involved in radio, either as professionals or as amateur experimenters, and to specially organised groups sitting round loudspeakers. It is safe to say that no other broadcast concert has ever made such an impact.

Notwithstanding the successes so far attained, there was no relaxation of the official attitude towards broadcasting. Service opposition hardened and the broadcasting of entertainment was stigmatised as the frivolous use of a national asset. This culminated in November 1920 in the banning by the Post Office of any further broadcasts from Chelmsford, a clampdown which coincided with the opening of KDKA and the launching of full-scale broadcasting in the United States. From this point on the Post Office found itself under pressure from three directions. First, there was a large body of amateur experimenters who wanted to be granted increased facilities, many of them demanding to know why it was that they were forced to listen to foreign broadcasts, such as the Sunday concerts regularly put out by The Hague, because there was no British service. Secondly, a growing number of ordinary people wanted to listen to broadcasting merely for pleasure. This group was particularly unwelcome in official circles as any suggestion that people should be allowed to use radio for entertainment was for some reason anathema. Finally, there were the manufacturers with their unemployed capacity, particularly the electric lamp manufacturers. One of these, British Thompson-Houston, at a meeting of the Wireless Society of London (shortly to become the Radio Society of Great Britain) had already demonstrated a commercial receiver capable of receiving concerts from The Hague. (It cost £30, quite a lot of money for those days.)

Eventually the Post Office was obliged to give way and from January 1922 the Marconi Company was allowed to resume its broadcasts, which it did from a station at Writtle, near Chelmsford, though it was still decreed that broadcasting must stop for three minutes in every ten so that an operator could check that there was no interference with any official transmission. The complete fatuousness of this condition, so reminiscent of the Red

Flag Act of the 1890s, is shown by the fact that not once during the whole period during which it was in force was the broadcasting station asked to remain silent. It is amusing to note that only four years later the boot was on the other foot when the postmaster-general was asked in the House of Commons whether steps could be taken to stop ships using spark transmitters near the coast, as they were interfering with broadcast reception.

Accounts of pre-BBC broadcasting often mention 2MT, Writtle and 2LO, Marconi House, London. There were, however, two other stations which must be mentioned. Early in 1922, A. P. M. Fleming, head of research with Metropolitan-Vickers (now Associated Electrical Industries), one of the companies which had been manufacturing valves during the war, set up a 50W transmitter at the company's laboratories in Trafford Park, Manchester. From this station, 2ZY, exercises and transmissions were broadcast from 16 May 1922. Their success led Metropolitan-Vickers to associate itself with the Radio Communication Co, a company which had been set up with the object of breaking the Marconi Company's monopoly of marine wireless and the two companies made a joint application to the postmaster-general for permission to erect two permanent broadcasting stations, in Manchester and in Slough. The Western Electric Co was also in the field, setting up a 500W transmitter at its office in Norfolk Street, London, only a few yards from Marconi House. This was shortly afterwards moved to Birmingham, where its first experimental transmission there as 5IT went on the air just in time to be able to broadcast the general election results on the evening of 15 November 1922.

When it became clear that the government would have to allow broadcasting, and that the question now was what form it would take, the Post Office at first favoured a system which would not give a monopoly to any one company. However, it soon became obvious that if such a solution were to be adopted the allocation of the limited number of wavelengths available to the various companies wishing to broadcast would present a most difficult problem. In April 1922 a meeting was held of representatives of interested parties, which included not only the Post Office and the manufacturing companies but also the armed services, the Foreign, Colonial and India Offices, and the Board of Trade. It was at this meeting that the Post Office put forward the suggestion that the companies wishing to engage in broadcasting should join together as a single company for the purpose. Of those present the postmaster-general claimed precedence.

What justification in law did the postmaster-general have for claiming the power to control wireless telephony? He based his claim on the Telegraph Act of 1869, which laid down that the postmaster-general had the exclusive right to transmit telegrams in the United Kingdom, and the 1904 Wireless Telegraph Act. As neither of the acts made any reference to telephony, particularly radio telephony, it is hardly surprising that the validity of his claim was questioned. In particular, Captain Wedgwood Benn raised the matter in the House of Commons, when he asserted that the acts mentioned did not give the postmaster-general the powers he claimed. However, the matter was never tested in the courts.

Meanwhile the public was clamouring for a regular broadcasting service and eventually the postmaster-general called together representatives of all the firms engaged in manufacturing radio equipment, with a view to working out a practical and acceptable scheme; on 18 May 1922 representatives of the six major companies involved, together with those of twelve smaller companies, met on neutral ground at the General Post Office in London. The manufacturers were afraid of losing business through delay and meetings followed each other in quick succession.

At an early stage it became clear that the ownership of patents would play a major part in the discussions when Mr Godfrey Isaacs, managing director of the Marconi Company, announced that his company owned a very large number of patents and that it would be impossible to build an efficient transmitter without infringing any of them. He also made it clear that if it was decided to allow other companies besides Marconi to erect and operate stations, it would not allow such companies to use its patents. The company favoured single control and it is difficult to see how, with the patent position as it was, anything else was possible.

It was not only the building of stations that was affected by the patent situation. The manufacturers were mainly interested in launching a broadcasting service because they expected that it would earn for them good dividends from the sale of receivers and components but here again the Marconi Company was in a commanding position, Mr Isaacs again stating that they owned a considerable number of patents covering the manufacture of valve receivers. Later this was disputed by the Western Electric Co, owners of the patent on the superheterodyne receiver, but this had not yet come into general use and the manufacturers had no option but to pay royalties to the Marconi Company.

For several years after the launching of broadcasting in Britain a large proportion of the receivers used were home-made. One of the advantages of making a receiver at home was that it avoided the payment of royalties, which were fixed at 12s 6d (62½p) per valve-holder for amateurs. But this of course was illegal and amateur constructors were informed that they must pay the same royalty as the manufacturers. History does not record how many actually did so! Another matter which involved the committee in much discussion was the amount of the licence fee to be paid by listeners, and the allocation of the proceeds. (One member of the committee expressed the opinion that the cost of providing the actual programme would be negligible.) Suggested charges for licences ranged from 5s (25p) to 15s (75p); eventually 10s (50p) was agreed upon, half to go to the Post Office, half to the BBC.

The British Broadcasting Company Ltd was formally incorporated at a meeting held on 18 October 1922 at the Institute of Electrical Engineers, representatives of some two hundred manufacturing companies being present. The composition of the board is interesting, comprising as it did only representatives of the Post Office and the manufacturing companies; none had any connection with the activities with which the new company would be concerned, namely the dissemination of news and the preparation of and dissemination of material designed for entertainment, education and religion. The Post Office representative was there mainly as a watchdog, protecting vested interests; the others were concerned with providing the necessary electrical circuitry. Very soon another non-member of the entertainment world was appointed general manager: J. C. W. Reith, a Scotsman, an engineer and a son of the manse. In his case, though, it can fairly be said that he represented religion, though not officially. It is, of course, a commonplace for directors to have no acquaintance with the technicalities of the industries they control but this is not usually the case for the whole board.

The first BBC station to be opened, 2LO London, sited on top of Marconi House in the Strand, went on the air on Tuesday 14 November 1922. The programme was confined to a talk by Sir John Noble, chairman of the company, in which he outlined the BBC's aims and policy. It would be most interesting to know the details of the first normal programme which was broadcast on the following day but there is no record of it, the earliest known programme being the one broadcast on Thursday 16 November. It consisted of a one-hour popular concert, 1900–2000hr. The following were the artistes:

Leonard Clarke	*Baritone*
Bruce Maclay	*Flute*
Guy Dowell	*Tenor*
Lily Clarke	*Contralto*
Stanton Jefferson, ARCM	*Pianist*

The first Sunday broadcast took place on 19 November. It consisted of a similar concert in two parts, 2000–2100 and 2130–2200hr. The first BBC news bulletin went out on 23 December 1922.

The Birmingham station was opened on 15 November, one

day after London. Though not the first station, its opening programme achieved a notable 'first' with its broadcasting of election results in Britain.

Manchester 2ZY was the next, a few days later, and it was not long before stations at Cardiff, Bournemouth, Newcastle, Glasgow and Aberdeen were added. However, these did not cover the whole country and it was found necessary to erect low-power relay stations at Plymouth, Leeds, Bradford, Dundee and elsewhere to fill in the gaps. All these operated in the medium waveband. Later, in 1924, a high-power station operating on 1500m (200kHz) and capable of covering the whole country was opened at Daventry.

The history of the birth and early days of broadcasting is taken up not so much with the surmounting of engineering problems as with the struggle against vested interests, the first of which, blocking the way to an adequate broadcasting service, was the armed services. They were able to raise cries of 'defence', 'national security' and the rest, which were valid enough in wartime but now, with army wireless barely ticking over and the navy severely cut down, could not be taken seriously. The second and perhaps the most serious obstacle was the reluctance, inherent in officialdom, to give anyone permission to do anything. This took the form first of an attempt to prohibit broadcasting altogether, and then, when that position became untenable, to make sure that if broadcasting could not be prevented altogether at least there must be as little of it as possible.

Today British broadcasting only shuts down for three out of the twenty-four hours and on special occasions not at all; but when broadcasting began it was only allowed between the hours of 1700 and midnight. It is difficult to see what justification there could have been for such a restriction. If broadcasting was to be permissible between those hours, if indeed it was to be permitted at all, it is hard to see why it should not have been permissible at any other time. At first the revenue from licences would not have been sufficient to pay for much longer hours; nevertheless, that was something which time would rectify and it was certainly no justification for writing a clause limiting the hours of broadcasting into the BBC's licence. In the event, the extension of hours was very slow and it was not until 1972, the Golden Jubilee Year of British broadcasting, that the last restriction on permitted hours was lifted. One can only think that the restriction stemmed from the Puritan tradition, very noticeable in early British broadcasting, that too much enjoyment was not good for one. In 1924 Reith expressed the opinion that there was no demand for lunchtime broadcasting and it does not seem to have occurred to anyone that housewives might like to listen to broadcasting at ten o'clock in the morning as they went about their work.

The group most likely to feel the impact of the new service was the one composed of the various sections of the entertainment industry and the theatrical managers certainly looked upon broadcasting with disfavour. Indeed, it is easy to see that the broadcasting of a new play running in the West End of London might lose a good number of potential theatregoers. The managers insisted that there must be no broadcasting of current plays in their entirety but they did permit excerpts likely to arouse interest in the play. There was, however, little opposition to the broadcasting of light entertainment, as it was quickly found that a few appearances in front of the microphone soon raised the box-office value of an artiste.

It was confidently predicted that broadcasting would kill the gramophone. We now know that it has had the reverse effect. As for serious music, at first the quality of reception was much inferior to what it is today and musical people were often scornful of broadcast concerts. However, technical advances eventually caught up and broadcast music became acceptable, which meant that good music was made available to large numbers of people who previously had had little opportunity to appreciate it.

Broadcasts to schools began in April 1924. One might have expected that teachers would resent the introduction of the loudspeaker into the classroom, perhaps regarding it as a criticism of their own ability to present a subject adequately, but in fact this did not happen. Teachers usually took the view that broadcast talks were supplementary to the ordinary classroom lesson, which indeed is what they were designed to be and not an alternative to it. There were two fields of education in which broadcasting proved itself particularly successful: one was in the teaching of languages, where it was possible to employ teachers with the correct accent, something not possible in every school; the other was in music, in the teaching of which sound is such an important factor.

One extremely formidable group with a considerable vested interest was the one which claimed a monopoly right in the collection and dissemination of news—the newspapers and news agencies. At a very early stage a government spokesman had declared that there must be no broadcasting of news that had not already been printed, apparently considering that the government had a duty to protect the interests of the press, and when it came to drawing up agreements the press and agencies showed that they were determined to oppose any incursion into their territory.

Today we have come to expect instant news as normal. Even if we cannot be shown an event actually happening on our TV screens, we are accustomed to having it reported to us immediately, even if it is something taking place on the moon. News bulletins are broadcast hourly, and if some big news breaks between bulletins we do not have to wait even until the next one. Not so in 1922; nobody at that date visualised such instant communication. Great efforts were made to get newspapers containing racing and football results on the streets within an hour or two of the events taking place, and some sudden sensation would sometimes warrant a special edition, but for the bulk of the news one had to wait for the evening paper in town and next morning's paper in the country. When World War I broke out, for instance, there was no way of informing the whole nation instantaneously and many people first knew of it when they opened the following morning's papers.

It was clear to those interested in starting broadcasting in in Britain that they could do much better than this, for they had in their hands an ideal instrument for the dissemination of news, as had already been demonstrated. As far back as 1910 amateur broadcasters in the United States had included news in their programmes; and when commercial stations opened there in 1920 and 1921 the press raised no objection to the broadcasting of news. The newspapers, in fact, took an active part in the launching of American broadcasting and by 1922 they owned between them eleven broadcasting stations. They themselves broadcast news, usually in the form of summaries designed to whet the listener's appetite and induce him to buy the paper in order to have the full story. Indeed, broadcasting itself was regarded as being newsworthy, and many American papers started radio columns or even sections. It was not until later that their attitude towards broadcasting changed. In the early days the volume of radio advertising was not significant but as broadcasting developed it grew rapidly and by 1935 it had made serious inroads into newspaper advertising revenue; and it was this that eventually caused the hostility of the press to radio in the United States.

In Britain the press, which included the London and provincial newspapers and the news collecting agencies, was from the outset hostile to the broadcasting of news. It had never hesitated to use wireless for the collection of news, and as we have seen it was a newspaper that made the first commercial use of wireless, but its use for the dissemination of news to the public was another matter. Corporately it had a monopoly in the field and it claimed the right to retain it. It was the old story of the vested interest opposing the intruder with new methods. To make things more difficult for the BBC the press had a powerful ally in the Post Office, which received a substantial part of its revenue from press telegrams and, not being able to see into the future, did not realise how large would be its share of the revenue from broadcasting licences. The press was particularly hostile to the broadcasting of sporting results and of running commentaries on important events

—in other words, outside broadcasts. These were from the first an important feature of American broadcasting.

A committee was appointed to draw up an agreement concerning the broadcasting of news at which it was eventually agreed that the news agencies should supply the BBC with news in exchange for an agreed royalty while on their part the BBC agreed that each news bulletin should be preceded by the words 'Copyright by Reuter's, the Press Association, Exchange Telegraph and Central News', and that none should be broadcast before 1900hr. The BBC's managing director, J. C. W. Reith, was quite agreeable to the late start. In his *Broadcast over Britain* he wrote: 'I do not believe that there is much demand for an earlier bulletin. A comparatively small proportion of people are in a position to listen before 7 pm.'

Tom Driberg, in his *Beaverbrook*, commented on the attitude of the press to the broadcasting of news:

> The popular newspapers were right in feeling acute alarm when the BBC came into existence and began to broadcast news and comment. They were wrong in supposing that it would prevent people buying newspapers. It was more subtly and fundamentally dangerous than that; by seeking to maintain certain standards of impartiality and objectivity, it has daily over the years taught people to stop *believing* newspapers, at any rate of the more garish sort.

Reith himself attacked the partiality and consequent unreliability of the press when, commenting on the restriction under which the BBC was allowed to broadcast a speech live but was forbidden to comment on it, he wrote, 'the account in one part of the press may be slanted favourably to a speech, the other part may almost ignore it'.

The first BBC news bulletin was broadcast at 1900hr on 23 December 1922. Reith tried to get the agreement with the press modified to enable it to broadcast commentaries on important ceremonies, sporting events and other occasions, but without much success. All the press would agree to was descriptions of the general scene but not the event itself. For instance, if the BBC wanted to describe the Derby, it could broadcast the noise of the crowd and the sound of the horses' hooves but not the race itself, and despite all the pressure he was able to bring to bear on the press the restriction remained in force until the end of 1926. It was, however, suspended for one important and significant occasion, the General Strike.

The government had finally been pressurised into allowing broadcasting, albeit only during limited hours. Clearly, however, they could not 'in the public interest'—whatever that may mean—allow just anything to be included in the programmes. It had sided with the press in the matter of news. What about controversy? There seems to have been a large degree of agreement among those negotiating the BBC licence that there should be no controversy at all, and considering that there was no such ban in the press, nor indeed in any other field, the prohibition cannot but be seen as most unreasonable, clearly indicating a desire to impose censorship. Kingsley Martin, commenting on it, wrote, 'anything that provoked thought was regarded as subversive by those of authority', which probably explains the reason for the ban accurately enough. At any rate it was imposed and rigidly enforced.

However, the matter was not settled for good. During the General Strike of May 1926 the BBC started to broadcast 'editorials' to take the place of those normally to be found in newspapers, and these were continued after the strike had ended. Their tone was conciliatory and they urged moderation; nevertheless, they did not please some members of the government and some of the BBC governors also disapproved, with the result that the postmaster-general intervened, insisting that the editorials must be submitted to him before being broadcast. As this proved to be impracticable, the editorials had to be dropped.

But the defences against controversy had been cracked and it was not long before attempts were being made by influential people to have exceptions made to the no-controversy rule for special purposes. When the corporation replaced the company

Reith was in a stronger position, and he put all the pressure he could on the Post Office to lift the ban. Finally he succeeded and on 5 March 1928 the ban was lifted, 'as an experiment'.

While negotiations were going on to decide what the public should or should not be allowed to hear, the public itself was steadily acquiring a taste for 'the wireless', its increasing popularity following technical advances which gave improved quality and simplified the operation of receivers. At first there were few valve sets; most of the early listeners had only headphones and crystal receivers of the 'cat's-whisker' type, which had a habit of going out of adjustment at particularly interesting points in the programme.

The crystal sets soon began to be replaced by valve receivers, usually home-made to specifications published in the new periodicals devoted to wireless. They were battery operated and so required an accumulator which had to be charged once a week, if the set was much used, and a dry battery of 100V or so, which tended to be expensive. Accumulator acid sometimes spills or creeps, and dry batteries, if neglected, do not always remain dry, with the result that the battery compartment of many cabinets became decidedly messy. Tuning, too, was sometimes tricky. There were usually two, sometimes three, separate variable condensers which had to be lined up, and often there was a separate control to each valve. To make things even more complicated, on many receivers wave-changing was not performed by turning a single switch but by changing sets of three plug-in coils. Having six knobs on the panel to twiddle might be fun for some member of the family mainly interested in picking up distant stations, but not for those who just wanted to listen to the programme. Perhaps the worst feature of those primitive receivers was the reaction, or feedback, which was included in most circuits. If the coupling was not properly adjusted the set would burst into oscillation, distorting the programme and causing interference with reception by the neighbours. Time, however, brought improvements. The introduction of the superheterodyne circuit, of valves that could be operated from the mains, and above all of one-knob tuning, not only greatly improved reception but made operating much simpler, and by the mid-1930s receivers, though still very bulky by present-day standards, had reached a reasonable standard of performance.

In 1925 a committee under the chairmanship of Lord Crawford was set up to consider the future of broadcasting, and in the following year it published its report which recommended the conversion of the British Broadcasting Company into the British Broadcasting Corporation, that is to say, a specially chartered body. But before the government could do anything about it the General Strike occurred. This was to have a profound effect upon both the official and the public attitude towards broadcasting, particularly the broadcasting of news.

The strike had for some time been casting its shadow before it and the government had made extensive preparations for dealing with it. From the outset they treated it as a national emergency, as did the establishment generally. Then, on the eve of the strike, the printers employed by the *Daily Mail* were given to print for the next issue a leading article headed 'For King and Country' which from the strikers' point of view was inflammatory in the extreme. They refused to print it and as a result work on practically all newspapers throughout the country ceased. It was at this point that broadcasting as a means of disseminating news really came into its own. The BBC could clearly no longer be fettered by restrictions on its right to broadcast news before 1900hr and it promptly appointed additional staff to its news department; it instituted a service of five bulletins a day, the first at 1000hr, containing not only straight news but government announcements and emergency transport arrangements. There were a few attempts by strikers to jam BBC transmissions but they proved to be abortive, as were attempts to dissuade strikers from listening to the broadcasts.

Soon after the start of the strike the government began publication of a daily news sheet, the *British Gazette*, to which the Trades Union Council countered with the *British Worker*. Besides these, a few papers managed to appear in skeleton form but the total number of papers available was still comparatively very small and the BBC's bulletins remained the principal source of news throughout the strike. By the time it was all over they had acquired a high reputation for accuracy and impartiality, though whether this reputation was entirely justified is not so obvious today as many things not generally known at the time have long since come to light. However, that is another story. One thing at least emerged very clearly: broadcasting was no longer just an interesting hobby nor merely a source of entertainment for the uncritical. It had become a power in the land.

With the reappearance of the newspapers the BBC reverted to its normal news service but it was now quite clear that the old restrictions on the broadcasting of news would have to be substantially modified. The Crawford Committee recommended that the company should become the British Broadcasting Corporation as from 1 January 1927 and a new agreement with the press negotiated with the time of the first news broadcast advanced to 1830hr. It was further advanced to 1800hr in September of that year. The BBC was also allowed to broadcast running commentaries on sporting and other events and the first commentary on a Rugby football match went out on 15 January 1927, on an Association football cup tie on 29 January, and on the Grand National Steeplechase in March.

With the relaxation of the restrictions on the broadcasting of news and of outside broadcasting came a liberalisation of the attitude towards controversy. Something like the present position was reached in 1931 when, before the General Election, party leaders were for the first time given time on the air. One of those who spoke was Sir John Simon, who recorded in his memoirs that it was one of the few occasions on which a speaker was placed before a microphone, free to say anything he wished without having first to submit his script for approval.

Words and Music: Part Two

The General Strike had shown for the first time, at any rate as far as British broadcasting was concerned, what a powerful—indeed dangerous—weapon had been placed in the hands of governments or, in fact, of any body that could gain control of a country's broadcasting system, and since then one of the first steps taken by any subversive group attempting to overthrow a government has been to seize the radio stations.

Control of its broadcasting stations not only gives the government the ability to speak to the people directly and so influence them in its favour, it also enables them to speak to the people of neighbouring countries with the same object. At first this was accomplished by the simple method of increasing the power of existing stations, so that they could be heard abroad, but as technology advanced it became the practice to erect special stations which could beam their transmissions in the direction of the intended recipient.

Such tactics provoked counter-measures and totalitarian regimes decreed severe penalties for those caught listening to foreign programmes. For instance, in Germany in 1938 the penalty for passing on news received from a foreign station was five years' imprisonment, while to make it difficult to commit this offence the 'people's wireless set' was made to receive only German programmes. Furthermore, while it was *verboten* to listen to the wrong programmes, steps were taken to ensure that everyone listened to the right ones. The country was divided into a thousand districts, each with a *Funküberwachung*, whose duty it was to see that when community or, more accurately, 'required listening' was ordered, every public square, factory and school was fitted with a receiver. Russia developed a system similar to the radio-relay system familiar in this country. Householders were allowed only loudspeakers connected by line to a central receiver, which of course only took the officially approved transmission.

As the prospect of another world war became more ominous, and the need for preparation more urgent, the volume of foreign-directed propaganda rapidly increased. As early as 1933 Germany was stepping up her English-language broadcasts, directed at British overseas territories and intended to impress their inhabitants in Germany's favour, and Italy began broadcasting to the Middle East. To counter this the BBC began broadcasting in Arabic early in 1938, at which time it had no German or French transmissions, and began broadcasting news intended for the continent of Europe in French, German and Italian at the time of the Munich crisis in 1938.

After the outbreak of war the volume of foreign propaganda broadcasting, together with the number of languages used, increased enormously and there was no easing off of this trend with the termination of hostilities. Table 6 shows the extent to which it had developed by the early 1950s. Unfortunately we have nothing to give us any idea of the effect of this vast outpouring or words—and of financial resources. To what extent were the objects of the broadcasters achieved?

TABLE 6

Broadcasts in foreign languages

Source	Number of languages used
BBC	46
Voice of America	38
Italian Diaro	34
Vatican City	25
Radio India (to Asia)	12

One cannot help feeling that all along those responsible for this class of broadcasting have been over-optimistic. It is one thing to speak, it is quite another to compel people, over whom one has no control, to listen. For many years propaganda broadcasts have been directed at Britain (together with other English-speaking countries) from both Russia and the United States. No doubt some British communists and fellow-travellers listen to those from Moscow, and obtain inspiration and encouragement from them, but this must make up a very tiny minority of the population. And as for the Voice of America broadcasts, it is doubtful if many people in Britain have even heard of them, let alone listened to them regularly. The only foreign propaganda broadcasts that ever had much of an audience in Britain were those of 'Lord Haw-Haw' during World War II. They were very cleverly conceived, and they did impress some people with the accuracy of the German Intelligence Service, but most people treated them as a joke.

However, not all foreign broadcasts have a sinister intention behind them. Some are intended to keep people living abroad in touch with their homeland. The BBC news bulletins are much valued by those living in remote places abroad and also by small newspapers in poor countries. In India, for instance, many small papers copy the radio news and print it as it stands.

If broadcasting can be used by autocratic rulers to consolidate their position, it can also provide a means by which all who wish to can hear the voice of a loved and respected sovereign. On 24 April 1924 King George V opened the Wembley exhibition with a speech that was broadcast over the whole country; never previously in this country, and probably never in any other since the days of the small city-states, had it been possible for more than a small fraction of the people to hear the sovereign's voice. Now it had been heard by vast numbers and from that date on it would be familiar not only to a handful of those near the throne but to nearly everyone in the country. Broadcasting had forged a new link between the Crown and the People such as had not previously existed. And by Christmas Day 1932 advancing technology had made it possible for King George V to be heard not only by his subjects in Britain but by those in the dominions throughout the world. In delivering the first Royal Christmas Broadcast he instituted a custom which, with modifications to suit changing circumstances, is still observed.

In 1926 the General Strike had shown what broadcasting, with its power of instant communication, could do in a national emergency and in the 1930s it had further opportunities to show its value in moments of high drama. The first of these occurred on the evening of 19 January 1936 when King George V lay dying. There are still many people who can remember the solemn reading of the bulletin which told the nation of the approaching end of his life, 'The king's life is moving peacefully towards its close.' Then, shortly after midnight, the inevitable announcement came, read by Sir John Reith. The king had passed away. Many kings and queens had died before that but never had there been such a feeling that the whole nation was at the bedside, something that no newspaper announcement could have given.

The second of those occasions in which broadcasting was dramatically involved concerned the ending of the brief reign of King Edward VIII which took place on 11 December 1936 and involved two major broadcasts. For several days the constitutional crisis involving the king had been developing, with public tension mounting hourly until the atmosphere was like that preceding a severe thunderstorm. Everybody was thinking of one thing. Everyone waited. At last, late in the afternoon, it became known that an announcement was to be made at 1700hr. People gathered round their loudspeakers and at that hour learnt that for the first time in their history a British king had abdicated.

The second broadcast took place late in the evening. There have been many occasions on which men in high places have had to make personal statements justifying their actions or perhaps excusing them; none, though, has had the poignancy, the sense of high drama as the one made that evening by the man who had risen from his bed that morning, still His Majesty King Edward VIII, King by the Grace of God of Great Britain and Northern Ireland and of the British Dominions beyond the sea, Defender of the Faith, Emperor of India, and who was now introduced to the listeners by Sir John Reith, Director-General of the BBC, as His Royal Highness Prince Edward.

The next major broadcast during the period was that of the coronation of King George VI and Queen Elizabeth. It was the first time that a coronation had been broadcast and this time it was a happy and joyous occasion. But the note was soon to change again. On 22 February 1938 Adolph Hitler broadcast a three-hour long speech in which he warned other nations of Germany's power and intentions. It might be described as the use of radio for frightening one's neighbours.

12
Television

So far we have been concerned mainly with the ability which the electro-magnetic waves gave us to transmit, first signals, and then all kinds of sound, with some small attention to their use for controlling mechanisms at a distance. We must now say something about their ability to add vision to sound. The idea that what we would now call closed-circuit television might be possible goes back to the middle of the last century, decades before Hertz had demonstrated the existence of the electro-magnetic waves which were to make broadcasting possible, and many more decades before the appearance of the devices that would be needed for a practicable television system.

The first step along the road to modern television was taken in 1873 when Willoughby and Smith discovered the photo-electric properties of selenium. This material is no longer used, having been superseded by more sensitive photo-electric substances, but experiments with it showed that it was possible to convert light into electrical impulses.

The second step was the invention, simultaneously by W. E. Sawyer in the United States and Maurice Leblanc in France, of scanning. At first this was carried out by the scanning disc, which came from Germany in 1884, but a lot more was needed before anything that could be called television was evolved, and progress was slow. Another development was that before the mechanical system of scanning had got very far, a rival appeared.

In 1911 a paper was read before the Röntgen Society of London describing a system using a cathode-ray tube that incorporated all the essential features of modern television. This became known as electronic scanning. It was not developed further at the time, because there were as yet no suitable amplifiers available.

In July 1915 the *Wireless World* printed its first article on television, in which the author was far from optimistic about its prospects. He expressed the opinion that in the near future there would be a system capable of transmitting pictures over the ordinary landlines but he could not see the possibility of transmitting them by wireless because 'to construct a wireless apparatus capable of transmitting and receiving 40,000 signals in a tenth of a second and arrange them in their correct order would tax to the utmost the powers of our cleverest inventors and prove the limit of human ingenuity'.

It was John L. Baird who first brought television to the notice of the public in Great Britain and who is still remembered by many as its great pioneer. He first attracted attention with a demonstration which he gave in Selfridge's department store in London and from that point he seems, at any rate on paper, to have progressed rapidly. On 4 September 1926 he demonstrated a system of stereoscopic television to the British Association for the Advancement of Science and in the same year he opened the first experimental television station in Britain, at Harrow 2TV. In the next year he moved to new headquarters in the South Tower of the Crystal Palace in South London, with which many people will associate him. He set up a company to develop his inventions, which manufactured sets of parts of television receivers for assembly at home, and these were on sale at Selfridge's for about £10. For advertising he used the slogan 'See the world from your fireside', which was a little premature, one may think.

In 1930 Dr J. Ambrose Fleming was asked to report on the Baird system. He wrote that in his opinion no other system could satisfactorily do the same as Baird's, but that most if not all its elements were old; what Baird had done was to make a workable whole of them. He had seen a picture of a single face transmitted so that it was possible to recognise it.

Meanwhile, in the United States, Dr Vladimir K. Zworykin, a scientist of Russian origin, had in 1923 demonstrated for Westinghouse a practical though still primitive electronic television system. At the same time experiments with the mechanical system were proceeding in America. In 1924 RCA transmitted across the Atlantic a picture of Charles Evans Hughes. This in effect was a single frame of a television picture. They developed this and in 1926 opened a commercial transatlantic picture service, transmitting photographs, printed matter and music scores.

Interest was growing, and in 1928 the magazine *Television* made its appearance in New York. In that year the first television play went out. It was entitled 'The Queen's Messenger' and it was performed before a fixed camera in the GEC experimental station, Schenectady W2XAD. Sound was broadcast from WGY and the receiver screen measured 3in by 4in. In the following year Dr Zworykin, now assistant director of the RCA laboratories, patented the Ionoscope, described as 'the eye of the television camera', and the Kinescope, a receiver screen. By 1935 interest had grown to such an extent that RCA allocated $1 million for television development.

In Britain, in 1929 the BBC made its first experimental TV transmission while in the same year the Marconi-EMI electronic system was demonstrated.

As the dream of a commercial TV service began to show signs of becoming a reality, rivalry between the mechanical and the electronic systems intensified, the exponents of each striving for the great prize, the adoption of their system for British television, for it was realised that there would be no room for an alternative system. In 1935 the BBC set up an experimental station at the Alexandra Palace, ideally situated on top of a hill in north London. In it were two studios, one for Baird and one for Marconi-EMI, and comparative tests were carried out. In February 1936 the fateful decision was made, with the BBC choosing the Marconi-EMI electronic system. The Baird mechanical system was out.

For Baird, who had been working on his system for many years, this defeat was final. Curiously, his popular image survived it, and people who had no knowledge of the technology of television continued to regard him as its inventor. As late as November 1953 an article about him in the *Sunday Mail* of Salisbury, Rhodesia, described him as 'the genius who gave us television', and even today he is referred to as the inventor of television, though, as with wireless telegraphy, no one person is entitled to that distinction. Actually the modern TV set incorporates little if anything derived from his work.

The author remembers hearing of the BBC's decision, and thinking what a shame it was that once again the labour of the small man with limited resources had gone for nothing and the prize had gone to the wealthy corporation. This, however, was not the cause of Baird's failure. When he embarked on his work the basic principles of both systems were known; but he chose the one which we can now see had no chance of final success. After his defeat he devoted himself to manufacturing TV receivers.

The decision having been taken, events moved rapidly, and on 2 November 1936 the BBC opened, from Alexandra Palace, the first regular television service in the world. At first it was visualised that TV programmes would consist of ordinary radio programmes, plus a picture of the artistes standing in front of the microphone, with a background of curtains, but such presentations did not last long. Scenery was introduced and then the microphone disappeared from the screen. Whole plays were televised, though not from the studio; as with the early radio plays, which came direct from the theatre, a camera was set up in the stalls and a performance then shot, a practice that lasted into the 1950s.

Outside televising of important events soon followed, the coronation of King George VI and Queen Elizabeth being one of the first. Then, in the early part of 1939, the Boat Race, the Derby, and Wimbledon appeared on British screens for the first time, but this was as far as prewar TV was allowed to progress. On 1 September, with the outbreak of war imminent, British television came to an abrupt halt. By this time there were about twenty thousand receivers in the country, all in the south-east.

Sir Harold Nicolson, in his *Diaries and Letters 1930–39*, describes this early television:

> ... I went round to the Beale's, where there is a television set supplied by the local radio merchant. We saw a Mickey Mouse, a play, and a Gaumont-British film. Compared with a film it was a flickering, dim, unfocussed, interruptable thing, about the size of a quarto sheet of paper. But as an invention it was tremendous and may alter the whole basis of democracy.

The manner of its closing down is reminiscent of the treatment of radio amateurs in 1914. The programme for 1 September 1939 was cancelled without any announcement. Viewers who switched on at the advertised time found themselves facing a blank screen and it was more than a month before any announcement regarding the closure appeared in the *Radio Times*. It is difficult to see what possible reason there can have been for such precipitate official action. No doubt the word 'security' would have appeared in any explanation, but it would have been impossible to have received the transmissions outside this country and it would not have mattered if they had been picked up. One can only suggest that it was suppression for suppression's sake.

Over to America again, RCA demonstrated television at the World's Fair in 1939, in the presence of President Roosevelt, who became the first chief executive to appear on viewers' screens. At the same time David Sarnoff announced the birth of a new industry—television. After the World's Fair, NBC instituted regular transmissions and on 1 July 1940 it opened the first regular commercial television service in America with fifteen hours of TV weekly. At first, however, the public did not show much interest, and RCA sold only four hundred receivers in five months. The American Telephone & Telegraph Co, in fact, took the view that the principal field for TV would be for personal use as an adjunct to the ordinary telephone, allowing documents and other things to be shown.

Then on 7 December 1941 CBS put out the first news telecast with news of the attack on Pearl Harbor. With America in the war, the Federal Communications Commission in 1942 prohibited the building of any more TV stations. The existing stations were not closed down but were mainly devoted to telecasting programmes dealing with such matters as civil defence and Red Cross training.

13
World War II

During World War II both sides were faced with a difficult problem: how to obtain the maximum advantage from the use of radio without at the same time letting it be of assistance to the enemy; the defensive measures employed to protect one's own use of radio, together with the offensive action taken against the enemy's radio, may almost be said to have turned the ether into a separate war zone.

Way back in 1904 the Russians had shown how serious could be the consequences of the careless use of WT, and ever since army authorities had regarded it with disfavour. In the military field the possibility of interception, with the consequent leakage of vital information, was still in 1939 a factor which had to be considered. And now a new and complicating factor had to be considered, one which not only affected the military use of radio but also broadcasting. Aircraft had come to depend almost entirely on radio aids for navigation, one of these being the device which enabled an aircraft to 'home' on any radio transmission that persisted for more than a very short time.

Early in 1939 broadcasting stations were situated all over Britain and with the possibility of war looming ahead it became clear to those in authority that they constituted a serious danger. But by the late 1930s the value of radio as a means of communication with the public had been established; the government wanted to make the fullest use of it if war should break out and the wholesale closing of stations could only be contemplated as a last resort. In July 1938 it was decided that the medium wave stations should be formed into two groups operating on only two wavelengths, 449·1m (668kHz) and 391·1m (890kHz). With several transmitters working simultaneously on the same wavelength it would be impossible for enemy aircraft to get a fix on any at a distance of more than a few miles. In addition, arrangements were made by which any station could be switched off within a few seconds should the RAF consider it a danger. If a station did have to go off the air, listeners would continue to receive the programme audibly from another point.

This system worked very satisfactorily throughout the war, in contrast to the German practice of switching off all stations during a raid.

Wartime radio in Britain began with the most momentous broadcast ever heard by British listeners with the announcement by the prime minister, Neville Chamberlain, at 1115hr on 3 September 1939. Many people will remember the occasion only too well. His tone, as he made his announcement, was solemn and full of emotion. Sir Harold Nicolson describes how he heard the broadcast at the home of a friend in the West End of London. His host had no radio set, but the housemaid had one, and just after eleven she brought it into the room where they were sitting. It seems strange that as late as 1939 only the servant in a well-to-do household had a radio receiver. It showed, in fact, that in certain circles radio had still not been accepted. It was something for the masses; intelligent people did not listen to it.

The war soon swept that attitude away. A radio set at least capable of receiving the news became a necessity of life. Loudspeakers even appeared in exclusive West End clubs, and one was actually installed in a committee-room of the House of Commons.

In contrast to the official attitude in 1914, there was now no attempt made to prevent people owning and using radio receivers; in fact it would have been impossible to have enforced any such prohibition, had it been thought desirable, and it was not. On the contrary, the government was most anxious that everyone should have access to a receiver, so that it could maintain direct and immediate contact with the public. Furthermore, 'the wireless' did much to maintain morale and, not least important, it enabled everyone to hear the inspiring orations of Mr Winston Churchill.

The restriction which had limited the hours during which the BBC was allowed to broadcast news was lifted for the second time, and lifted for good.

Broadcasting had completely revolutionised the dissemination of war news. During the Napoleonic wars it was the mail-coaches which spread news of victory or disaster. They carried the few newspapers for the literate, and at every coaching-inn along the main routes the coachmen passed the news by word of mouth, to be spread locally by horsemen and carrier's cart. From the time when despatches reached London it might be several days before their news reached distant parts of the country. The coming of the railways and the telegraph lines speeded things up considerably, but right into World War I one day's news did not reach the majority of people until they opened their newspapers on the following morning.

By 1939 radio had brought in the era of instant news. Immediately important news was received it was broadcast, so that it could be heard immediately by everyone . . . or perhaps we should say, by everyone whose batteries had not run down. Many receivers still required accumulators that needed charging and bulky 100–120V high-tension batteries. With increased demand and shortage of materials, the supply of the latter became very difficult, although the government did everything possible to maintain it.

The BBC's newscasts became of great importance not only to people at home, but to many in the occupied countries. They acquired a very high reputation for reliability, and many people, desperate to know what was really happening in the world, risked heavy penalties by listening to them.

World War II

Also of great importance was the broadcasting of the chimes of Big Ben. Every evening they were heard not only by listeners in homes and shelters in Britain, but in cellars and attics all over occupied Europe, where little bands gathered round secret radios, bands of people who were doing everything they could to harass the invader. To them it was a reminder that Britain was still there and free, and that hope was not dead.

So important were the chimes that it was considered too risky to chance them being put out of action by enemy aircraft. They were therefore not broadcast live but from a recording. This had the additional advantage that their absence would not indicate to the enemy that the Houses of Parliament had been hit.

On the outbreak of war the BBC found it necessary to rearrange its services. On the one hand it had to free some of its channels for service use; on the other it had to provide entertainment for the armed services—both for those training at home and for the British Expeditionary Force in France. In January 1940 a BBC official went over to France to assess the troops' requirements and was asked, 'Why can't we have music to listen to while we're dressing?' As a result the BBC started broadcasting to the forces at 0615hr. Less than twenty years earlier Reith had expressed the opinion that there was no demand for lunch-time broadcasting. Another group that had to be catered for was the munition workers; for them the BBC started 'Music While You Work', two half-hour sessions of popular instrumental music, 'the vocal' being dropped soon after the start because it was found that some workers were wasting time writing down the words.

Wartime radio was not only a booster of morale; it also helped to bring home to those not actively engaged in the struggle the reality of war in a manner and with a vividness that had not been possible in any previous conflict. This was never better exemplified than in what is sometimes referred to as the Gardner broadcast. At 0915hr on 14 July 1940, the BBC broadcast live a shot-by-shot, bomb-by-bomb running commentary on an aerial battle taking place over the Straits of Dover.

On that summer morning Charles Gardner, the BBC's ace war correspondent with the RAF, had the extraordinary luck to find himself with a recording van on a hill overlooking the Straits of Dover just at the right moment. He was standing watching a convoy steaming through the straits when a formidable force of 'bandits', comprising 40 Junkers 87 bombers with a strong escort of Messerschmidt 109 fighters appeared out of the east and proceeded to dive-bomb the convoy. At that moment a squadron of RAF Spitfires roared in from the opposite direction and immediately an aerial battle was joined. Gardner immediately contacted London and the programme then going out was interrupted to take his commentary live. Never before had a radio audience listened to anything like it. Through the sounds of the firing and the cheers of the onlookers near the van they heard a vivid description of the battle overhead.

Besides supplying news and entertainment for those at home, the BBC performed other vital services, one of the most important being the support of and communication with resistance fighters and the population generally in the occupied territories. Frank W. Silla, in his *The Silent War*, describes how the people of Guernsey, living under German occupation, were encouraged by the BBC broadcasts. At first they were allowed to retain their receivers, but when the Germans saw how they reacted to what they heard they decreed that all receivers must be surrendered. At once there sprang up an underground industry turning out crystal receivers; at the same time the earpieces in many telephone kiosks mysteriously disappeared. So great was the urge to have reliable news that a little band of enthusiasts copied the news and actually produced a clandestine daily newspaper which they passed round. Two of them were caught and executed. And the desire for authentic news was not confined to the Channel Islanders. Silla records that in the later stages of the war there were German soldiers who, tired of Nazi propaganda, used the receivers that had been seized to listen to the BBC news.

The BBC's European transmissions did not only contain news.

Buried in them were code phrases, apparently quite innocent, which carried vital information or instructions to resistance fighters in many parts of Europe.

During the early part of the war the owners of American broadcasting stations did not spend much on the gathering of war news, as it was not popular with sponsors, who feared that it might offend some sections of the public. To fill the deficiency the BBC from 1942 onwards produced a series of programmes entitled 'Britain to America' for broadcasting in the United States over the NBC network.

The organisation of broadcasting to bring instant news of the progress of a battle reached its peak in the preparations for keeping the people of America informed about the D-Day landings. With American forces now very much involved, the priority of war news had risen considerably, and for this great event all the American networks pooled their resources. Their men among the hundred war correspondents aboard the ships and with the troops sent to London, where a centre had been set up, on-the-spot accounts of what was happening. Here Edward Murrow and his assistants edited them and compiled a programme which was transmitted to America on a US Signal Corps channel. The broadcasts started going out from all networks in the United States at 0330 EST. During the twenty-four hours after the news of the landing broke, CBS devoted seventeen hours to news from France. However, elaborate though this organisation was, it did not bring to the American people the first news of the landings. This came from the German Transoceanic News Agency.

In all the countries taking part in the conflict radio was considered a vital link between the government and the people. This is reflected in the way the number of receivers in use increased, as shown in Table 7 on the following page.

TABLE 7

Increase in the number of radio receivers

Country	1938	1946
	(thousands)	
Bulgaria	34	205
Czechoslovakia	1,000	1,616
France	4,166	5,577
Spain	300	400
Soviet Union	4,550	9,300
Total, all Europe	17,242	35,459

And by 1954 the number of radios in the world exceeded for the first time the number of copies of newspapers printed daily:

| Radios | 257 million |
| Newspapers | 256 million |

Radio was not only used by both sides for telling the world its own version of the news; it was also used for spreading doubt and uncertainty. The best known examples of this were the broadcasts of 'Lord Haw-Haw', whom we have already had occasion to mention. A potentially more dangerous tactic of the Germans was one which they employed early in 1945. In France the US forces on one section of the line had suffered a temporary reverse as a result of a surprise German attack but the situation was stabilised as a result of joint American-British action, as explained in a communique issued by Field-Marshal Montgomery. When the Germans broadcast on the BBC wavelength a distorted version of this communique, which made it appear that Montgomery claimed all the credit, considerable offence was given to the Americans and there was a certain amount of coolness at staff level until what had happened had been made clear.

Very few people in Britain, when they read of the heroic deeds of resistance groups in the occupied countries, had any idea that there might be, perhaps within sight of their own homes, a secret wireless station ready for use if need be by British resistance fighters. David Lampe, in *The Last Ditch*, describes the elaborate organisation which was set up to deal with the worst of all possible eventualities, the establishment by an invading force of a bridgehead on British soil. Groups called Resistance Units were formed to make up what might be termed a shadow resistance force. Such groups would need communications and for this purpose a two-tier system of radio stations was set up. One tier consisted of a number of stations dotted about the country, set up and working openly. But each had attached to it a number of clandestine UHF stations, usually underground and concealed with great ingenuity; most of them were manned by specially selected members of the ATS.

For guiding bombers to their targets the Germans did not depend only on being able to 'home' on conveniently placed British transmitters. Those who remember the early air raids of World War II will recall that they were relatively small affairs, compared with what were to follow, with only a few bombs dropped, and it is now believed that the real purpose of these raids was to try out a new radio navigational system which the Luftwaffe was developing.

Basil Collier, in *The Defence of the United Kingdom*, tells how at the time of the fall of France in 1940, papers were found in a shot-down enemy aircraft indicating that the Germans possessed a new beam system for guiding bombers to their targets. From what were called *knickebein* stations, twin beams were transmitted, one consisting of a series of dots, the other of a series of dashes. Two such pairs could be made to cross over a target. The discovery was regarded as extremely serious and an aircraft was sent up to investigate. It succeeded in picking up one of the beams, which was found to pass over the Rolls Royce works at Derby. This resulted in the formation of RAF 80 Wing, a special force whose task was to combat and interfere with, by all possible means, the transmissions from the *knickebein* stations. One of the methods it employed was to pick up the beams and re-radiate them in a different direction, so that bomber pilots would not know which to follow. Another was direct jamming. Such measures were to a large extent successful but they did not entirely prevent the use of the system, which was employed in the bombing of Coventry.

As in World War I radio was of immense importance in the conduct of the war at sea. In the first place it was vital for the control of the British fleet, which was scattered over all the oceans; for this the Admiralty had made extensive preparations and when war broke out it possessed a network of stations at bases all over the world. Radio was also very much involved in the protection of merchant ships and the destruction of raiders. It could be used to give warning of the position and movements of submarines, and it enabled a ship being attacked to warn other ships of the presence of submarines in the area, something that the Germans did their best to prevent by knocking out the ship's wireless installation.

The Germans sometimes used radio with disastrous results for themselves. At one period they operated their submarines in groups called wolf-packs and on occasions the units would become separated. In order to reassemble them the parent ship would radiate homing signals, signals which were occasionally picked up by British warships or aircraft which used them to home on the submarines.

The keeping of radio silence was usually of great importance but sometimes the reverse was useful. Transmissions could be used to deceive the enemy. In *War at Sea 1939-45* S. W. Roskill describes how in 1942 WT was used to give false information to the enemy. At that time the British naval forces in the South Atlantic were stretched to their limit and were urgently awaiting reinforcements. To give the impression that these were on their way the naval station at Freetown, Sierra Leone, made a practice of communicating nightly with two imaginary ships, which had been given the call-signs of the *Inflexible* and *Invincible*, of Battle of the Falkland Islands fame. One wonders whether the enemy spotted the connection.

The British Army entered World War II a great deal better equipped in the matter of radio than it had been in World War I. Nevertheless the position was far from satisfactory. Because of the fear of interception, the use of radio was severely limited, with the result that, as had happened on numerous previous occasions, the operators did not get sufficient practice and so when they were needed in an emergency they were not as efficient as they should have been. Another old trouble which recurred was insufficient range. In the spring of 1940 the British Army found itself engaged in the German new form of fast-moving, mobile warfare based on large numbers of tanks and self-propelled guns. For communication in such conditions radio was essential but it was soon found that the sets supplied to the army had not been designed to cope with the ranges now demanded. The General Staff announced that they needed an RT set with a range of a hundred miles which could be placed inside a vehicle; no such set was available and the best that could be done was to improve the range of existing sets with better aerials and other modifications. It was not until 1944 that sets capable of the required performance became available.

In World War I the outstanding success of wireless as far as land warfare was concerned was its use as an aid to air-spotting for the artillery. In World War II it was in the area of long-distance communication.

For the second time, the news that the great struggle had ended reached the war-weary peoples of the world through the medium of radio. This time, however, it came not from the victorious commander-in-chief but from the defeated foreign minister. At 1405hr on 7 May 1945, Count Schwrin von Krosigk, German foreign minister newly appointed by Grossadmiral Dönitz, Hitler's successor, broadcast from Flensburg a statement to the effect that Germany had surrendered unconditionally. The allied leaders were not prepared for this announcement and, pending confirmation, they attempted to suppress the news; the attempt completely failed, for it had already spread round the world. The result was that by the time the Ministry of Information in London announced that tomorrow would be VE-Day, and Prime Minister Churchill had told the country over the radio that hostilities would cease at one minute after midnight, the streets were already filled with crowds celebrating victory.

14
The Eagle Has Landed

In the field of radio communication the coming of peace marked the approaching end of the era of the valve and the supremacy of sound broadcasting, and the beginning of the era of television and the semi-conductor.

For nearly half a century the valve and the semi-conductor (first known as the crystal) had existed side by side. The first silicon crystal detector was patented in Britain in 1904, the year of Fleming's valve, and the first carborundum detector in the United States in 1906, with various other crystals following in quick succession. The two systems therefore started level. The crystal detectors were used for communication work for about fifteen years before they were superseded by the valve, but for several years after that they were extensively used for broadcast receivers and they never faded completely out. However, their use was limited to detection and rectification; they could not amplify or oscillate.

In 1941 the junction diode semi-conductor appeared; this was made up of two crystals fused together. It was followed, in 1946, by the junction triode, developed by the Bell Telephone Laboratories in America. This consisted of three crystals fused together, which could be made to oscillate, and was called a transistor. Its technology was developed very rapidly and it soon began to displace the valve, being much less bulky and consuming less current.

The transistor made possible the age of miniaturisation. This had begun with the production of the miniature and then the sub-miniature valves for war purposes, but now it was possible to make radio apparatus of almost unbelievable smallness. In the domestic field the big table radios disappeared, to be replaced by small 'transistor' receivers; the 'personal' walkie-talkie trans-receiver also made its appearance.

Parallel with the advance of the transistor came the rebirth of television. In America, wartime restriction on television had not been so severe as in Britain, the main ban being on the manufacture of receivers. This ban was lifted in 1946 and on 17 September of that year the first postwar sets went on sale. From this point development was rapid, as is shown in Table 8. The popularity of TV was also spreading at a prodigious rate elsewhere and it has been estimated that by 1960 the number of sets throughout the world had reached the 225 million mark.

TABLE 8

Year	No of TV receivers
1949	1 million
1951	10 million
1958	50 million

In Britain the BBC was allowed to resume its television service in 1946 and from then on development was rapid. With larger screens and all-round improvement of quality its popularity grew and grew, and a new social function made its appearance, the TV party. This was an entertainment given by the owner of a television set for neighbours who were not so fortunate and comprised an evening of viewing and light refreshments. The new cult had many other social consequences, some to be regretted. It undoubtedly affected attendance at club functions and such activities as amateur drama; it also proved the death knell of many picture houses and was the cause of much domestic discord. It was a splendid form of relaxation for the tired businessman but it discouraged home-made entertainment.

A big boost was given to the spread of television by the announcement that the coronation of Queen Elizabeth II was to be televised, and dealers were inundated with orders for sets to be installed in time for the occasion. The televising was a great success, though unfortunately when the programme was shown in America it was sometimes associated rather too closely with advertisers' products, which gave some offence on both sides of the Atlantic.

From this point onwards television may be said to have replaced radio as the principal form of home entertainment. But it was not merely an entertainment. By adding vision to the spoken word it had greatly improved on the power of radio to transmit information; it had, in fact, become the most important of the communications media.

With the miniaturisation of components it became possible to make use of higher and higher frequencies (or shorter and shorter wavelengths). Apart from the advantage that they gave of being less congested, the use of these frequencies made possible the spread of radio communication into yet further fields. Low and medium frequency waves are reflected from the layer of ionised atmosphere which envelopes the earth, so they cannot pass into outer space; only a small fraction of VHF waves (30–300mHz) is reflected, the remainder shooting off tangentially into space. All of UHF (300–3,000mHz), the band used for modern TV, shoot off; and so do SHF (3,000–30,000mHz).

The equipping of the police seems to have proceeded very slowly when one considers how much it was needed. At one time the only means which some forces had of communicating with a car on patrol was by having an arrangement with certain garages by which, on receipt of phoned instructions, they would display a card. An officer on patrol would see this and stop to phone HQ. However, the use of higher frequencies, together with miniaturisation, greatly helped development.

Towards the end of the war a start was made with equipping the police with a VHF system. At first the sets were very bulky by present-day standards, consisting as they did of separate transmitting and receiving units powered by a motor generator whose current consumption from the car battery was very heavy. It was not until the early 1960s that the introduction of transistors made it possible to make substantial reductions in the size and current consumption of car transmitters.

With regard to 'walkie-talkie' equipment, we have seen that as far back as 1932 the Brighton-type receiver was introduced. After that there appears to have been little progress for a long period, and it was not until 1959 that the Home Office became seriously concerned in finding a really advanced instrument and approached various manufacturers who might be interested in meeting its requirements. In July 1961 the Technical Subcommittee of the Police Wireless Communications Committee decided on a specification that was issued as PWC 388. The weight of this apparatus was estimated at 2lb and the cost at £100. Following this, interest was centred on three new types of send-receive equipment, and large-scale experiments with them were begun in 1965. But before any conclusions could be drawn from these experiments, Messrs Pye introduced the Pocketfone UHF equipment. This was so satisfactory that further experiments with VHF equipment were discontinued.

Another field in which radio on a local scale has proved itself to be of immense value is in the Fire Service. We have noted that the Metropolitan Fire Brigade had two of its stations equipped with Marconi apparatus as early as 1900, but this seems to have been premature. Radio would have been of immense value in fighting the fires of World War II, but suitable apparatus was not available until nearly the end of the conflict. However, experiments were being made, and radio was first used between a mobile fire appliance and its control centre in 1940. A year or two later VHF equipment was introduced.

The great value of radio in fighting fire lies in its saving of time when time is most precious; it renders unnecessary the search for telephone boxes, and it enables the officer in charge of an appliance to phone for further help without stopping if he sees as he approaches a fire that it is bigger than he can cope with.

From the field or local radio we pass to a field that is very far from local. The most spectacular advance in radio communication during the last two decades has been the part it has played in the development of space travel. As UHF and SHF waves pass straight out into space, it is possible to use them for communicating with a satellite or spacecraft. Such communication is, of course, absolutely necessary in order to control a space vehicle from earth, and without some means of sending information back to earth unmanned vehicles would be useless.

Back on earth, the development of VHF and still higher frequencies has revolutionised long-distance communication. From an early date in its history WT was used for communicating over very rough country where the cost of installing and maintaining landlines would have been excessive. Such terrain is usually found in the tropics, where the medium and long waves used in the early days are often plagued with almost continuous static; VHF is not troubled by this but its surface range does not extend much beyond the horizon. This difficulty can be overcome by employing a chain of repeater stations and many tropical countries, such as Nigeria, now possess a network of VHF stations linked by relays, which also provide far cheaper and more reliable WT and RT services than would be possible with landlines. The relay stations are usually spaced at twenty-five mile intervals; they are unmanned, and only need to be visited periodically for refuelling.

Although relay stations cannot be erected over long stretches of ocean, the use of VHF and still higher frequencies has made possible a revolution in intercontinental communication. As these frequencies are radiated tangentially out into space, they can be picked up by a receiver suitably situated far above the earth's surface, in other words on an earth satellite, and from there they can be re-transmitted back to earth at the desired point (Fig 8). Such a system will transmit high-quality speech and television pictures between widely spaced points on the Earth's surface.

The primary cost is heavy but an SHF carrier wave can be used and this can carry more than a thousand telephone conversations simultaneously.

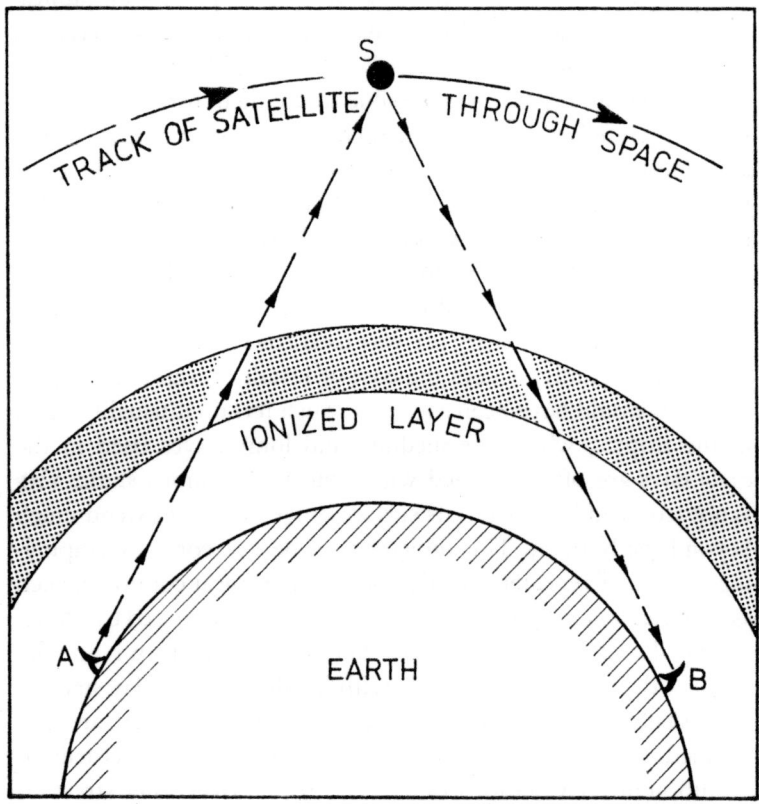

Fig 8 Long-distance transmission via satellite. As the Earth rotates the position of S between A and B remains unchanged

The first communications satellite, Telstar 1, was launched from an American base on 10 July 1962. It was in the form of a sphere, $34\frac{1}{2}$in in diameter and weighing 170lb; it contained 16,000 parts and was powered by a 15W battery charged by 3,600 solar cells. An attempt was made, between 2126hr and 2326hr on that same day, to transmit a programme from the United States to Britain, but without success. The following evening the same programme, which consisted of talks by President Kennedy and other officials in Washington, was successfully transmitted to Britain, the route being by landline from Washington DC to Andover, thence via Telstar to Goonhilly Down, Cornwall. A separate transmission was made to Pleumeur-Boden, France. That same evening transmissions were made in the opposite direction, in the form of a test card from England and a short programme of songs from France. At that time transmissions via Telstar 1 were restricted because the satellite was orbiting the earth at a comparatively low altitude, about a hundred miles, that is to say out of step with the surface, so that it was only in a position from which relaying was possible for a short period during each orbit. This difficulty was overcome by placing a satellite at a height of 2,300 miles from the earth. At that height the speed of a satellite is equal to the speed of rotation of the surface of the earth, so that it appears to remain stationary over a given point and relaying can take place at any time.

In the few years since the launching of Telstar 1 there have been tremendous advances in the use of communication satellites and in 1964 the International Communications Consortium was formed to provide satellite channels for the transmission of WT, RT and TV, the first commercial communications satellite, Intelsat 1, being launched in April 1965. These advances have not made the beam stations obsolete, nor are they likely to do so. Whatever their technical advantages, satellite stations are expensive and if they become faulty, as did Intelsat 2, they cannot be repaired. To be profitable they must carry heavy traffic all the time; nor can they be regarded as immune from destruction in time of war. In comparison with them the HF beam stations are as safe from attack as any system could be.

We set out to examine the ways in which the exploitation of the electromagnetic waves has enabled man to increase his control over his environment, and we have seen that this has been achieved to an extent which nobody engaged in its early development could possibly have foreseen. It has made life at sea infinitely less dangerous; it has provided greatly improved means of communication between points on the earth's surface; it has made modern aviation possible; it has brought cheap entertainment to vast numbers of people; and now it has enabled man to venture into outer space. It is a fine record by any standard, though not without blemish. It would be possible to have filled this book with accounts of the endless disputes over radio patent rights; during the first three decades or so of radio communication vast amounts of money went into the pockets of lawyers hired to show that the idea contained in one patent did or did not cover the idea described in another patent; at the same time some large corporations did very well out of clever inventors who were naïve in business, and men whose work brought untold benefit to the community were rewarded with less than is commonly showered on some tawdry entertainer.

However, the day of the lone garage or garret inventor has now passed. New inventions, revolutionary developments, now come from expensively equipped laboratories financed by wealthy companies who have no option but to bid for the services of brilliant scientists with attractive salaries. It may not be so glamorous but it is more rewarding for the inventor.

We will close with a brief word about the most spectacular of man's technical achievements, an achievement in which radio played a major role: the first journey to the moon. This is not the end of the story of radio, of course, already radio is bringing us news of Mars, but we have got to end this book somewhere. It would be quite impossible to give here any useful account of the incredibly complex radio instrumentation that made this journey to the moon possible. Instead we will conclude by quoting the words of Neil Armstrong, spoken on 20 July 1969, the first words carried by radio from the surface of the moon to earth: 'Contact light on, engine off, the *Eagle* has landed.'

Appendix A
Particulars of Early Instruments

THE INDUCTION COIL

This instrument, sometimes referred to as a Rhumkorff coil, was used in laboratories as a source of high voltage. It was an essential feature of the early Marconi transmitters, and until about 1920 was used as a stand-by transmitter aboard many ships for use if power failed. It is shown in Fig 9. There is a core C made up of a bundle of soft iron wires; over this is wound a thick-wire

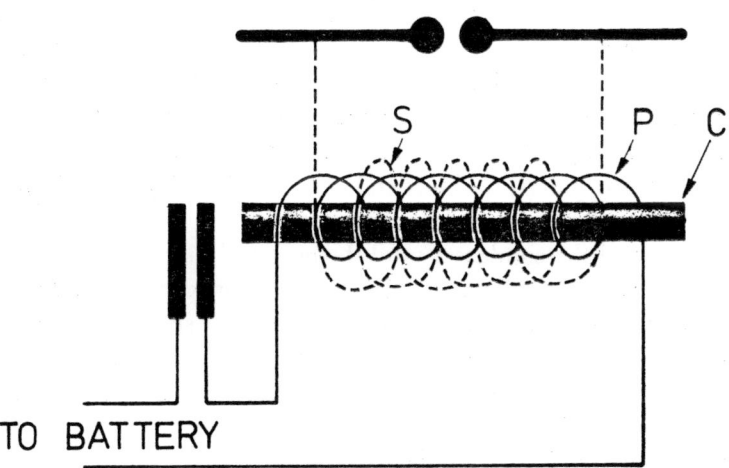

Fig 9 Induction coil

primary P and then a secondary S consisting of up to 50,000 turns of fine wire. Current from a battery passes through the primary circuit which includes a make-and-break as in a bell. The secondary output is taken to two brass rods mounted on ebonite pillars. The rods end in a pair of brass balls facing each other. These form the spark gap, the length of it being adjustable. In operation the make-and-break causes the primary current to rise and fall and so set up a current in the secondary winding. This is stepped up and when the potential difference between the balls reaches about 10,000V a spark occurs between them. One of these coils is shown on page 51.

THE LEYDEN JAR

This was an early form of condenser, consisting of a glass jar the inside and outside of which were each covered with metal foil to within a short distance of the top. Contacts were made with the inside and outside linings and taken to terminals. A bank of these jars could be built up into a condenser of any capacity or to take any voltage.

HERTZ'S APPARATUS

This is shown in Fig 10. A battery A supplied current to the primary of an induction coil C through key B. Across the secondary was connected a circuit comprising a Leyden jar condenser D and a spark-gap F. The latter consisted of two metal rods mounted on insulating pillars, the inner end of each rod having a metal ball, the outer end being connected to a copper plate P. The induction coil charged the condenser until the potential difference between the two balls was sufficiently high (10,000V or so) to break down the atmospheric resistance between them. A spark then leapt across the gap, allowing an oscillating current to surge backwards and forwards between the two plates until the condenser was discharged. This caused the plates to set up electromagnetic waves in the surrounding atmosphere and these in turn set up oscillations in the loop L, causing a minute spark across the gap.

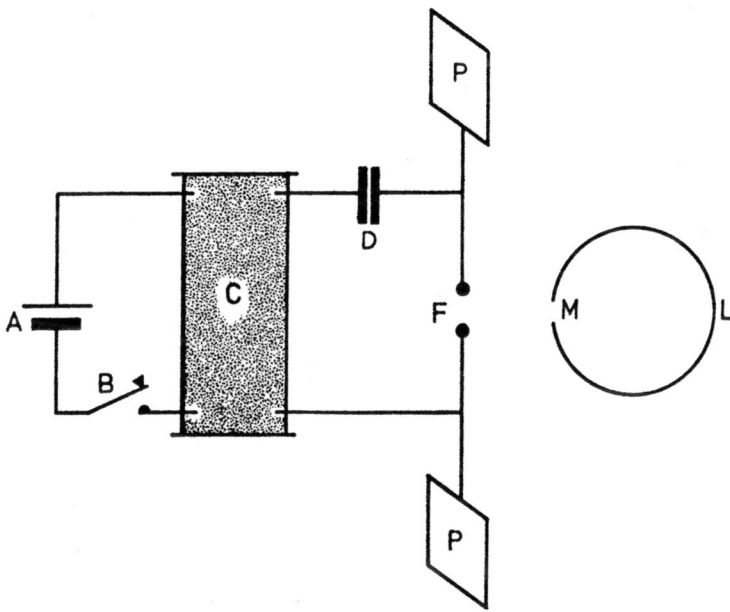

Fig 10 Hertz's transmitter and receiver

EARLY DEVELOPMENTS

Fig 11 shows Marconi's transmitter as it was when he landed in England. The circuit is the same as that used by Hertz except that the two radiating plates have been replaced by an elevated aerial E and an earth connection G. His receiver is shown in Fig 12 in which H is the coherer and J the Morse inker.

Marconi saw that the transmitter had a fundamental defect. The waves were radiated by one oscillating aerial–earth circuit and picked up by another. In order that there should be a maximum of power first radiated and then picked up it was necessary that in both these aerial–earth circuits the current should be able to surge up and down with the greatest possible freedom, that is to say the electrical resistance should be as low as possible. But the transmitting circuit included the spark gap, which had a very high resistance. Similarly, the aerial–earth circuit in the receiver

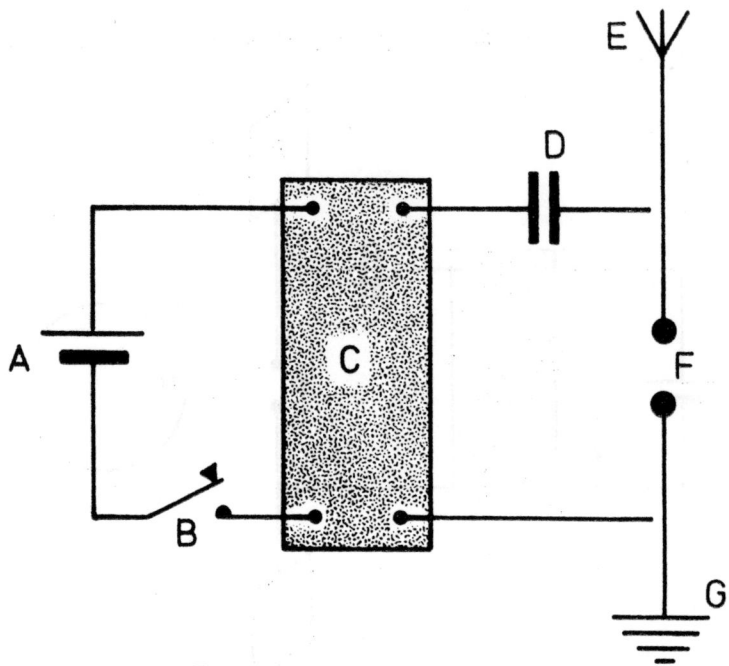

Fig 11   Marconi's first transmitter

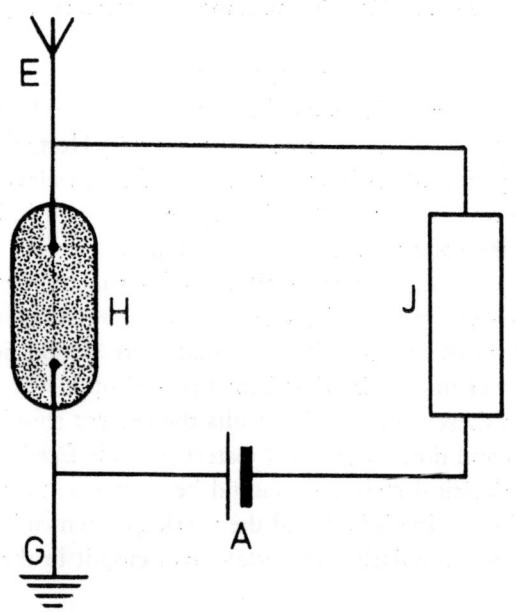

Fig 12   Marconi's first receiver

Particulars of Early Instruments

had the coherer in it, also with a high resistance. Marconi conceived the idea of forming a separate primary oscillatory circuit coupled to the aerial–earth circuit through a high-frequency transformer having a primary K of one turn and a secondary L of several turns of wire (Fig 13). In the same way the coherer was

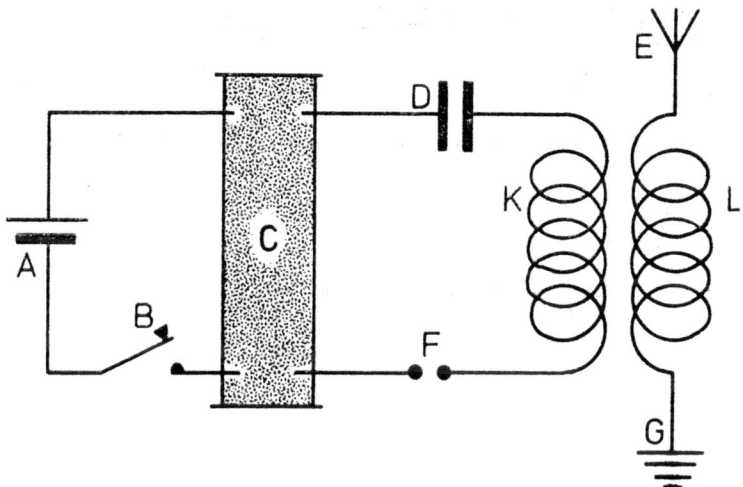

Fig 13 Improved Marconi transmitter

separated from the aerial–earth circuit by another transformer (Fig 14).

Meanwhile Lodge was also working on the same problem but his solution was to remove the spark gap from the aerial–earth circuit not by the inductive coupling of two separate coils, but by coupling the two circuits by means of a single coil common to both, that is to say by self-induction (Fig 15). He used the same device to couple the circuits in the receiver. This arrangement became a feature of the Lodge-Muirhead system.

SPARK GAPS

In the first transmitters oscillations were set up by causing a spark to occur across the gap between two fixed metal balls or spheres.

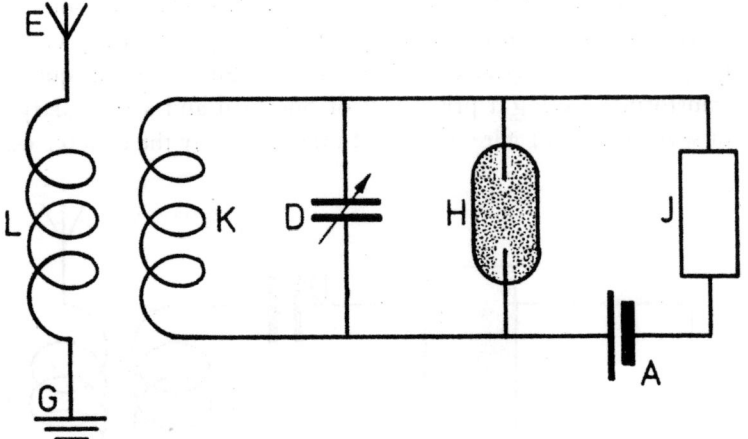

Fig 14 Improved Marconi receiver

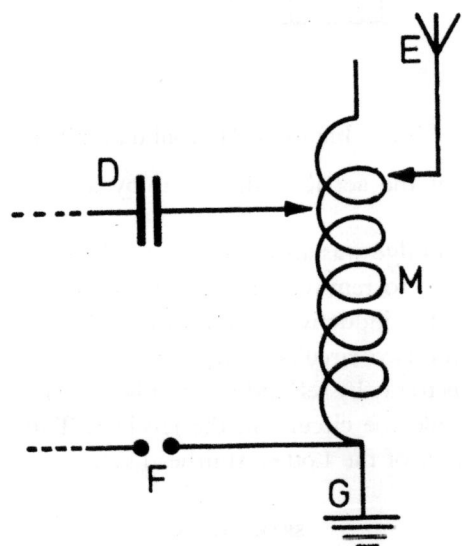

Fig 15 Lodge transmitter

Particulars of Early Instruments

These oscillations died away until they were insufficient to bridge the gap. Only the first few were of use and any beyond produced insufficient energy and so represented wasted power. It was, therefore, desirable to cut off the spark as quickly as possible.

One method of doing this was to use a rotary spark gap, which consisted of a toothed wheel which rotated rapidly between two fixed electrodes. When a pair of teeth passed close to the fixed electrodes a spark occurred. The rotation of the disc drew this out and cut it off. Another method was the Quenched Gap system. The gap consisted of a pair of thin metal plates with accurately ground faces and separated by a very thin mica washer. The plates had flanges which quickly dissipated the heat caused by the spark and so 'quenched' it or cut it off. Several such assemblies were fixed in a frame, so that there were a number of gaps in series. The number controlled the amount of power radiated.

THE RADIO FREQUENCY ALTERNATOR

The lowest frequency on which radio waves can be transmitted is about 15 kHz, which is well beyond the capacity of conventional alternators, and in practice much higher frequencies are needed.

In 1903 R. A. Fessenden designed the first alternator intended to put power into an aerial at radio frequency without the use of a spark discharge. Erik Alexanderson of GEC developed this further. In the conventional alternator designed for an output at 0·5 kHz the stator and field coils are arranged alternately in a ring and a toothed disc of soft iron is rotated beneath them. In the Fessenden-Alexanderson machine the stator and field coils are rotated at high speed in opposite directions, giving a much higher frequency.

Another successful machine was the Goldschmidt RF alternator. This consisted basically of two or three alternator units on one shaft. In Fig 16 the first unit comprises rotor coil R_1 and field coil F_1. The alternating output from R_1 is fed to F_2, and the much higher frequency output from R_2 may if desired be fed to a third unit to produce a still higher frequency.

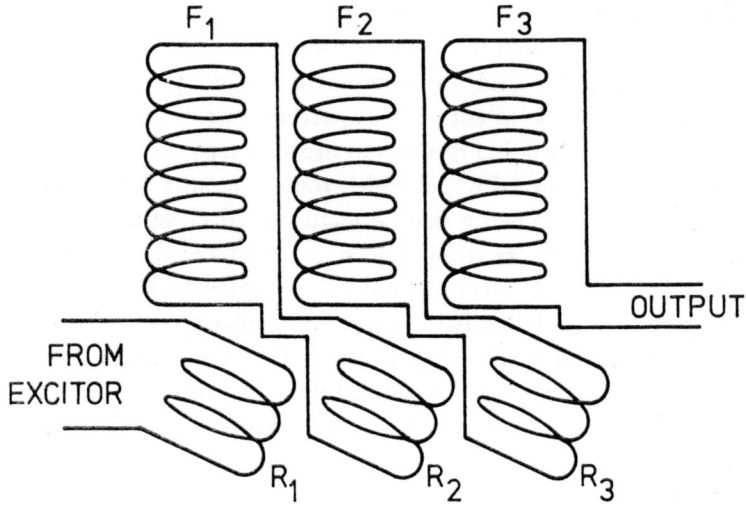

Fig 16 Goldschmidt radio frequency alternator

THE POULSEN ARC TRANSMITTER

The arc lamp was a form of electric lamp giving a very powerful light, in common use prior to World War I. Essentially it was made up of two plain carbon rods set in adjustable holders so that they touched end to end. A heavy direct current was passed through them; they were then drawn apart, with the result that the heavy current continued to flow across the intervening space, giving an intense white light. Such an arc did not behave according to Ohm's Law. Professor Duddell found that if he connected a large condenser across it, high-frequency oscillations were set up. If their frequency was within the range of audibility the arc produced a note. Because of this it was called 'the singing arc'. However, Duddell found that in practice it was impossible to make it produce oscillations of a frequency of any use in radio.

In 1903 a Danish engineer, Poulsen, replaced the positive carbon with a water-cooled copper rod. He then placed the assembly in a chamber filled with hydrogen gas and with a power-

Particulars of Early Instruments

ful magnetic field at right angles to the arc. An oscillatory circuit was connected across the arc, with the inductance connected to the aerial inductance. This arrangement could be used to transmit continuous oscillations of a useful frequency. As it was impossible to stop and start the arc at high speed, keying was performed by shorting a section of the inductance (Fig 17). This instrument was widely used in early CW and RT transmitters.

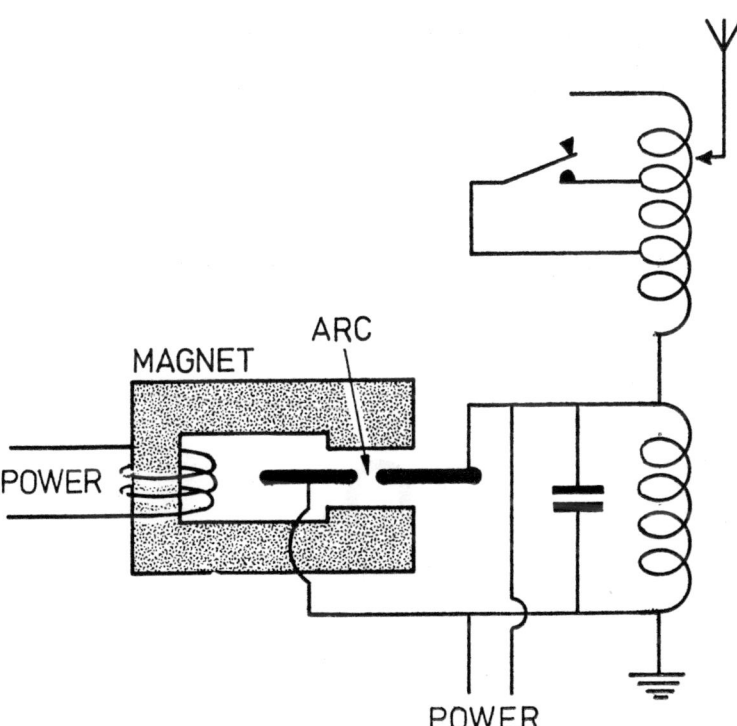

Fig 17 Poulsen arc transmitter

THE COHERER

The coherer consisted of a glass tube about 5cm long and 4mm in diameter in which were two small silver plugs, separated by a quantity of fine metal particles, F, loosely packed. Contact

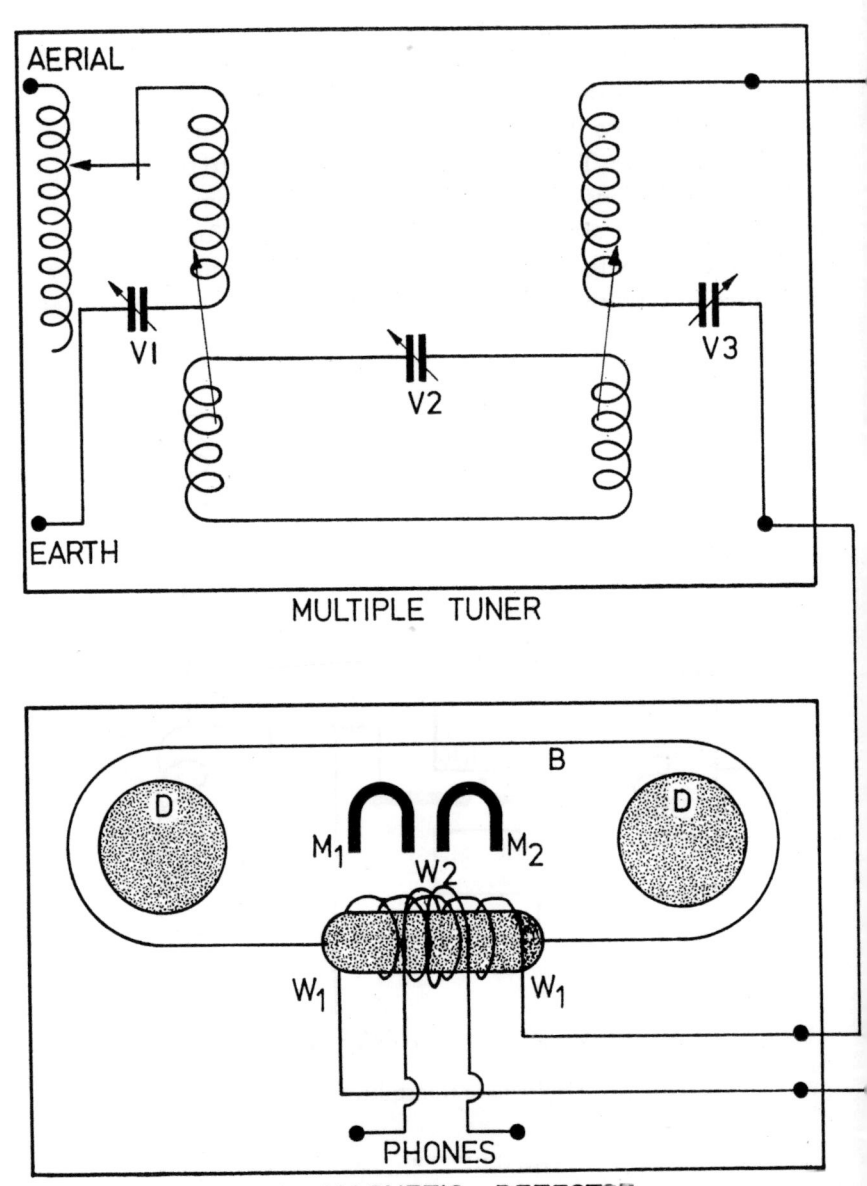

Fig 18 Multiple tuner and magnetic detector

Particulars of Early Instruments

with the plugs was made by wires sealed through the glass. One of these was connected to an aerial and the other through a battery and galvanometer to earth. When oscillations were received they caused the metal particles in the tube to coalesce or cohere which reduced their electrical resistance and so allowed current from the battery to flow through them and deflect the galvanometer needle, or operate a morse inker.

THE MULTIPLE TUNER AND MAGNETIC DETECTOR

These two instruments (Fig 18) made up the first receiving unit fully to meet the requirements of commercial WT. The tuner consisted of an aerial circuit, variably coupled to an intermediate tuning circuit, similarly coupled to a tuned output circuit. In operation it was necessary to first set variable condenser v_1 roughly and then using two hands adjust v_2 and v_3 simultaneously.

In the magnetic detector there was a short piece of tube over which was wound a primary w_1 of heavy wire, with over that a secondary w_2 of fine wire. Above this assembly were two permanent horseshoe magnets, M_1, M_2. A continuous band made up of fine soft iron wire, B, passed through the tube and over two pulleys D. A clockwork motor turned the pulleys and so dragged the band through the tube. The movement of the band dragged the field of the magnets with it. When oscillations passed through the primary winding they demagnetised the band. This caused the magnetic field to collapse and in doing so set up a uni-directional current in the secondary, with the result that each train of received oscillations caused a click in the phones.

This receiver was sensitive by the standards of the period; it was robust and reliable, and it gave good service for many years.

THE ELECTROLYTIC DETECTOR

This was developed in the United States by R. A. Fessenden (Fig 19), with versions by others. A glass tube contained an electrolyte E consisting of nitric acid. In this was immersed a platinum plate P and the tip of a fine platinum wire W. W was connected

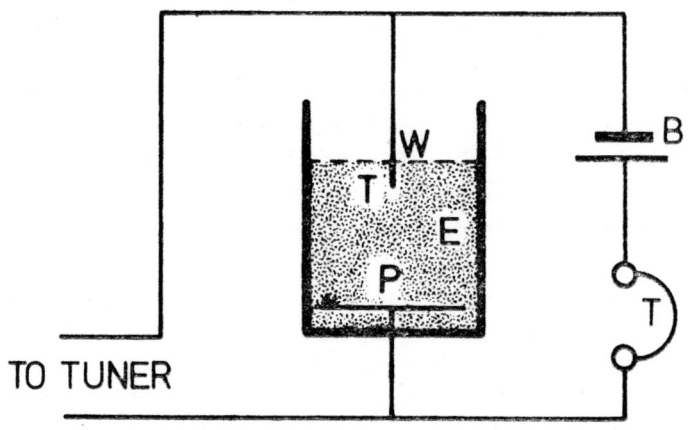

Fig 19 Electrolytic detector

to the positive pole of a battery, the negative pole being connected through a telephone T to P. W and P were also connected across a tuner with aerial and earth. When current passed through the circuit PEW the tip of W was quickly polarised (covered with hydrogen gas), greatly increasing the resistance of the circuit. An incoming train of oscillations from the tuner momentarily depolarised the tip, reducing the resistance and causing current to flow through T. Consequently a series of received trains set up a note in T.

THE PHYSIOLOGICAL DETECTOR

This strange instrument, described in J. A. Fleming's *Wonders of Wireless Telegraphy*, must be regarded as a scientific curiosity. In the eighteenth century the Italian scientist Galvani discovered that the nerve in the muscle of the leg of the common frog is very sensitive to electrical impulses. Professor Lefeuvre of the University of Rennes made use of this fact to construct what he called a physiological receiver (Fig 20). A frog was fastened to a board and a rod with a stylus was fixed to one leg. The output of a tuner was connected across the nerve of the leg. When an eletrcical impulse was passed through the nerve it caused the muscle to contract and this in turn caused the stylus to record the signal on a drum the surface of which had been covered with lamp-black. Lefeuvre actually received time signals from Paris on this instrument but it does not appear to have gone into commercial production.

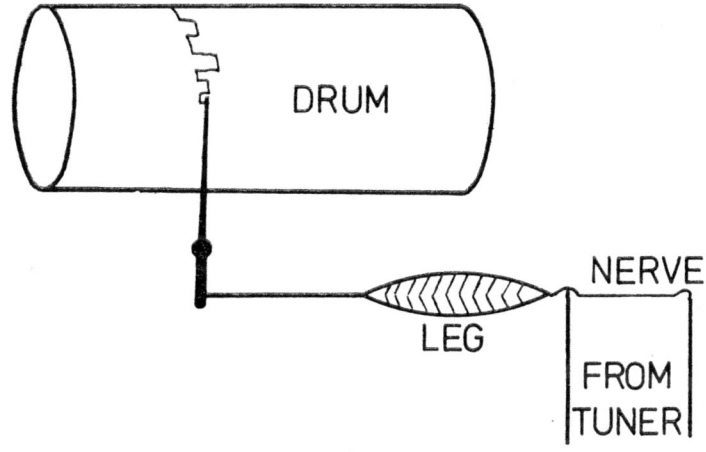

Fig 20 Physiological detector

WT IN EARLY MILITARY AIRCRAFT

The fitting of WT in early open-cockpit aircraft presented several difficulties. These included engine noise; the airscrew slipstream; the difficulty of preventing apparatus from being splashed with oil; bulk; weight; and above all the risk of sparks igniting aero spirit.

The transmitter was based on a 6in induction coil taking current from an 8 or 10V accumulator. There was a Leyden-jar condenser and an inductance of 14 SWG copper wire wound round a former. The output was fed to a pair of trailing wires, which acted as aerial and earth. The power was rated at 50–60W. Because of engine noise, normal audio reception was impossible and many early machines could only transmit. When they could receive, it was accomplished with a coherer which operated a battery and lamp circuit, so that the received signal was visual. A novel morse system was employed, a one-second flash for a dot and a two-second flash for a dash. It must have been exceedingly slow, probably not much more than two words per minute.

RUGBY RADIO

Call sign GBR. Opened in 1926.

Range: worldwide.

Aerial system: supported by 10 × 800ft masts, 8 for WT, 4 for RT (2 common).

WT power unit consisted of 54 × 10kW water-cooled valves, supplying 500kW to the aerial.

Frequency: 16kHz.

Total power used for WT, RT, and ancillary systems: 1,400kW.

RADIO-SONDE

This is a device used by meteorologists for measuring conditions in the upper atmosphere. It was first used in France in 1927, but was perfected a few years later by the Russian scientist Molchenoff. It consists of a small balloon specially designed to rise at an even rate (Fig 21), with below it a suspended parachute bearing instruments. The instrument readings are transmitted automatically to the ground by radio. When the balloon reaches a height of about 60,000ft the pressure of the expanding gas inside bursts it, and the instruments are returned to the ground by the parachute.

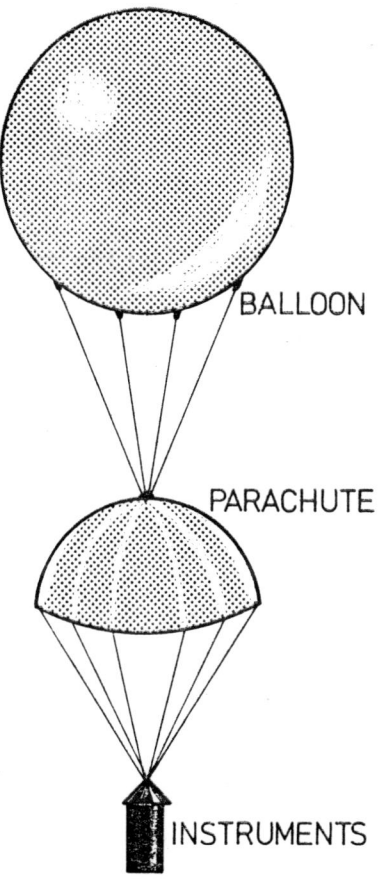

Fig 21 Radio-sonde system for meteorological observations in the higher atmosphere

Appendix B
Abbreviations

WT	Wireless telegraphy
RT	Radio telephony
SW	Short wave
DF	Direction finding
VHF	Very high frequency
UHF	Ultra high frequency
SHF	Super high frequency
kHz	kilo-Hertz (ex kilo-cycle)
mHz	mega-Hertz (ex mega-cycle)
BBC	British Broadcasting Corporation (or Company)
NBC	National Broadcasting Company
CBS	Columbia Broadcasting System
AT & T	American Telephone & Telegraph Company
RCA	Radio Corporation of America
RFC	Royal Flying Corps

The words 'wireless' and 'radio' are synonymous. At first the former was nearly always used; nowadays it is the latter. In this book 'wireless' usually goes with 'telegraphy' and 'radio' with 'telephony'.

Also synonymous are 'valve' (British) and 'vacuum tube' (American).

Appendix C
Chronological Table

1842	United States. *First* experiment with telegraphy without wires. S. F. B. Morse.
1845	J. B. Lindsay experiments across River Tay with conduction system.
1864	James Clark Maxwell shows theoretically that electromagnetic waves must exist.
1873	Photo-electric properties of selenium discovered.
1873	(20 August) United States. Lee De Forest born.
1874	(25 April) Guglielmo Marconi born.
1876	Telephone invented by Alexander Graham Bell.
1879	*First* transmission of music by landline.
1879	Principle of coherer discovered by D. E. Hughes.
1880	Scanning proposed by W. E. Bowyer in United States and Maurice Leblanc in France.
1882	United States. Graham Bell, using conduction method, communicates between two vessels $1\frac{1}{2}$ miles apart on River Potomac.
1882	*First* use of telephone for military purposes. Egyptian War.
1884	Scanning disc invented in Germany.
1887	Heinrich R. Hertz transmits and receives electromagnetic waves across a room.
1888	United States. Edison discovers that electronic particles are emitted from an electric lamp filament.
1890	J. A. Fleming begins investigating Edison effect.

1892	Sir William Crookes suggests Hertzian waves could be used for signalling.
1892	E. Branly and Sir Oliver Lodge invent coherer independently.
1894	Guglielmo Marconi begins his experiments in Italy.
1896	(6 February) Marconi moves to England. Granted first patent.
1897	(20 June) Marconi Wireless Telegraph & Signal Co formed.
1897	(November) *First* coast station opened at Alun Bay, Isle of Wight.
1898	Lodge patents tuned circuit.
1898	(20 July) *First* vessel fitted, the tug *Flying Huntress*.
1898	*First* commercial use of wireless.
1898	*First* sports commentary (of yacht races) given by Marconi from tug to Dublin *Daily Express*.
1898	(August) Osborne House and royal yacht linked to give news of Prince of Wales's progress after accident.
1898	*First* transatlantic liner fitted, *Kaiser Wilhelm der Grosse*.
1898	(December) East Goodwin Lightship and South Foreland Lighthouse linked.
1898	*First* calling device in operation.
1899	*First* British warships fitted, *Alexandra*, *Europa* and *Juno*.
1899	(October) United States. Marconi reports America's Cup races by WT.
1899	*First* United States warships fitted, *New York* and *Massachusetts*.
1899	(March) *First* used to summon help to ship in distress; East Goodwin Lightship.
1899	(October) *First* military use, the Boer War; Marconi Company send six sets to South Africa.
1899	(November) *First* shore-ship news bulletin; the *St Paul*.
1900	Two stations fitted for Metropolitan Fire Brigade.
1900	United States. *First* use of wireless in connection with meteorology.

1901 United States. De Forest Wireless Telegraph Co formed.
1901 (1 March) United States. Service between five Hawaiian islands opened.
1901 (21 May) *First* British liner, *Lake Champlain*, fitted.
1901 Marconi Company open coast stations at Crookhaven, Holyhead, North Foreland, Caister-on-Sea, Withernsea, Rosslare.
1901 (12 December) *First* transatlantic signals received by Marconi.
1902 Marconi introduces the magnetic detector.
1902 United States. *First* transmission of CW by Fessenden.
1902 United States. *First* broadcast of human voice by Fessenden.
1902 United States. *First* use of WT by Signal Corps.
1903 United States. Poulsen arc system introduced.
1903 *First* transatlantic Marconigram published in London *Times*.
1903 *First* international wireless telegraphy conference.
1904 J. A. Fleming patents diode valve.
1904 J. C. Bose patents semi-conductor crystal detector.
1904 *First* WT section of Royal Engineers formed.
1904 *First* Wireless Telegraphy Act passed by British Parliament.
1904 CQD distress signal introduced.
1904–5 Russo-Japanese War; both sides use WT.
1905 All ships of British and US fleets now fitted.
1906 (24 December) United States. *First* ever broadcast of programme of speech and music by Fessenden.
1906 Crystal detectors issued to Royal Navy.
1907 Two WT companies attached to cavalry.
1907 *First* WT signals received in air by Lieutenant C. J. Ashton in a balloon.
1907 Bellini-Tosi DF system introduced.
1907 United States. *First* use of RT to/from a mobile station; De Forest on Hudson River ferry.

1907	(17 October) Clifden–Glace Bay transatlantic service opened.
1908	*First* transmission from a balloon at Aldershot.
1908	De Forest broadcasts records from Eiffel Tower, Paris.
1909	(23 January) Liner *Republic* sunk in collision; help summoned by WT; 410 rescued.
1909	(29 December) Post Office take over Marconi Company's coast stations.
1909	United States. *First* political broadcast; De Forest's mother-in-law speaks on women's suffrage.
1910	(12 January) United States. De Forest broadcasts voice of Caruso from stage of Metropolitan Opera House.
1910	(17 July) United States. *First* communication with plane in flight; McCurdy at Sheephead Bay, New York.
1910	(29 September) *First* British plane-ground communication by actor Robert Loraine.
1910	*First* time signals from Eiffel Tower.
1911	*First* two-way communication in England with aeroplane.
1911	Imperial conference approves construction of imperial wireless chain.
1911	TV system with electronic scanning described before Röntgen Society of London.
1912	(15 April) *Titanic* disaster; *first* use of SOS signal.
1912	*First* use of WT by beleaguered garrison; Turkish Army at siege of Adrianople.
1912	Army manoeuvres; airship demonstrates value of air reconnaissance linked with WT.
1912	Marconi timed-spark system introduced.
1912	*First* ship fitted with DF; the *Mauretania*.
1912	United States. De Forest sells Audion patent to AT & T.
1912	United States. Congress passes first Radio Act.
1913	(October) Wireless Society of London (later Radio Society of Great Britain) formed.
1913	*First* use of WT by explorers; Mawson expedition to Antarctic.

1914	(3 August) Admiralty send WT signal to all ships and stations, 'Hostilities to start at midnight'.
1914	(4 August) Britain enters World War I.
1914	(September) *First* used in aircraft spotting for artillery; Battle of Loos.
1914	*First* air–ground RT in England.
1915	(July) *First* article on TV in *Wireless World*.
1915	*First* satisfactory RT sets supplied to RFC.
1916	(30 May) Movement of German High Sea Fleet detected by DF; Battle of Jutland.
1916	Marconi starts experiments with short-waves, leading to development of beam system.
1918	*First* direct transmission London–Australia by Marconi on SW WT.
1918	United States. Edward H. Armstrong patents superheterodyne circuit.
1918	DF used to give warning of approaching Zeppelin raiders.
1918	President Wilson's fourteen-point broadcast from Newark, New Jersey, with worldwide range.
1918	(11 November) Eiffel Tower broadcasts Marshal Foch's order for 'cease fire'.
1919	Successful use of WT and RT on transatlantic flight of airship R34.
1920	(23 February) *First* regular daily broadcasting service in world started by Marconi Company from Chelmsford, England.
1920	(23 February) *First* regular news broadcasts in world as above.
1920	(March) *First* flight of a civil airliner fitted with WT; London–Paris.
1920	(May) *First* commercial RT service in world opened; London–Madrid.
1920	(20 June) Dame Nellie Melba broadcasts from Chelmsford.

1920	United States. *First* advertisement broadcast by Frank Conrad from his garage in Pittsburg.
1920	United States. Westinghouse begins commercial production of receivers.
1920	(2 November) United States. *First* regular daily broadcasting service opened with presidential election results; KDKA Pittsburg.
1921	(2 January) United States. *First* religious broadcast; KDKA.
1921	(15 January) United States. *First* political broadcast by a commercial station; KDKA.
1921	(2 July) United States. *First* broadcast of a big fight: Dempsey–Carpentier; RCA station.
1921	United States. Detroit police start experimenting with use of radio.
1922	(16 May) Manchester 2ZY start experimental transmissions.
1922	(28 August) United States. *First* paid commercial broadcast by New York WEAF.
1922	(12 November) British Broadcasting Company formed.
1922	(14 November) *First* BBC station, London 2LO, opened.
1922	(15 November) *First* broadcast of general election results; opening programme of Birmingham 5IT.
1922	(23 December) *First* BBC news bulletin.
1923	Metropolitan Police start use of WT van.
1923	(1 May) BBC move to Savoy Hill.
1923	Successful test of wireless-controlled aeroplane; Etaples, France.
1923	(21 June) United States. *First* president (Harding) speaks on radio.
1923	United States. Vladimir C. Zwyrikin demonstrates practical electronic TV system.
1923	(2200hr, 23 August) *First* broadcast of a time signal by BBC
1924	(24 April) *First* broadcast by a British sovereign; King George V opens Wembley Exhibition.

Chronological Table

1924   (29 April) *First* direct SW RT; Poldhu–Sydney, Australia.
1924   United States. *First* transatlantic picture service opened by RCA.
1924   BBC start broadcasting to schools.
1925   United States. RCA introduce mains-operated radios.
1925   *First* British broadcast from an aeroplane.
1926   (May) General Strike.
1926   *First* flight Spain–South America with only DF navigation.
1926   (October) Britain–Canada beam service opened.
1926   *First* experimental TV station opened at Harrow 2TV by J. L. Baird.
1926   United States. National Broadcasting Company formed.
1927   Beam services from Britain to Australia, India and South Africa opened.
1927   British Broadcasting Company becomes Corporation.
1927   (15 January) *First* broadcast of a rugby football match.
1927   (29 January) *First* broadcast of an association football cup tie
1927   (March) *First* broadcast of the Grand National steeplechase.
1927   *First* experimental broadcast on SW from Chelmsford.
1927   United States. Columbia Broadcasting System formed.
1927   United States. *First* car radios manufactured.
1928   (5 March) Ban on controversy in BBC broadcasts lifted.
1928   Magazine *Television* appears in New York.
1929   British cable and wireless interests merged to form Cable & Wireless Ltd.
1929   Pedal wireless developed in Australia.
1929   (1 September) *First* experimental TV transmission by BBC.
1930   Baird TV products on the market.
1931   *First* broadcast by party leaders on eve of a British general election.
1932   *First* pocket receivers adopted by police.

1932	BBC's Empire Service opened.
1933	(March) United States. *First* fireside chat by President Roosevelt.
1934	United States. Song of nightingale from English wood broadcast live by CBS network.
1936	(January) Broadcast announcement of approaching death of King George V.
1936	(February) BBC adopt Marconi–EMI TV system.
1936	(2 November) *First* TV service in the world opened by BBC from Alexandra Palace, London.
1936	(11 December) BBC broadcast announcement of King Edward VIII's abdication and ex-king's farewell message.
1938	BBC start foreign-language broadcasts.
1939	*First* telecasts of boat race, Derby and Wimbledon.
1939	(30 April) United States. *First* appearance of chief executive on TV; President Roosevelt at World's Fair.
1939	United States. NBC start regular test telecasts.
1939	United States. *First* telecast of a baseball game: Columbia–Princeton.
1939	(1 September) BBC TV closed down.
1939	(3 September) At 1115hr prime minister broadcasts 'Britain at war with Germany'.
1940	(19 July) Live commentary of aerial battle by Charles Gardner.
1940	*First* use of radio between a mobile fire appliance and control centre.
1940	(1 July) United States. Commercial TV service opened by NBC from WBNT.
1941	United States. NBC demonstrate colour TV.
1941	*First* junction diode produced.
1941	(7 December) United States. *First* TV newscast from CBS; news of Pearl Harbor.
1942	United States. Ban on manufacture of TV stations and receivers.

1944	American Forces Network start broadcasting in Britain.
1944	(6 June) All US networks pool massive radio coverage of D-Day landings.
1945	(7 May) 1405hr; Count Scheverin von Krosingk broadcasts from Flensburg 'unconditional surrender of Germany'.
1946	BBC TV service resumed.
1946	United States. Ban on TV manufacture lifted.
1948	United States. *First* junction triode (transistor) announced by Bell Telephone Laboratories.
1951	*First* coast-to-coast TV network.
1951	(June) CBS start limited colour TV service.
1954	For first time number of radios in world exceed number of copies of newspapers printed daily.
1955	*First* BBC VHF FM station opened at Northam, Kent.
1957	*First* radio-controlled satellite launched by Russians.
1962	(10 July) United States. *First* communications satellite (Telstar) launched.
1962	(11 July) *First* TV programmes via Telstar exchanged between Europe and America.
1965	United States. *First* commercial satellite (Intelsat) launched.
1969	(20 July) *First* RT and TV transmissions from surface of the moon to earth.

Selected Bibliography

Barnouw, Erik. *A History of Broadcasting in the United States* (New York, 1960)
Briggs, Asa. *A History of Broadcasting in the United Kingdom* (3 vols) (1961–70)
Calder, Basil. *The Defence of the United Kingdom* (1950)
Chevainix-Trench, R. C. 'Wireless with the Cavalry', *The Cavalry Journal* (1929)
Churchill, Winston S. *World Crisis 1911–18* (1920)
Fleming, J. A. *The Wonders of Wireless Telegraphy* (1914)
Hancock, H. E. *Wireless at Sea* (Chelmsford, 1950)
Jones, R. A. *The War in the Air* (Oxford, 1923–31)
Kendrick, Alexandra. *Prime Time: The Life of Edward R. Murrow* (1970)
Liddell-Hart, B. *Memoirs* (1965)
Lodge, Sir Oliver. *Past Years* (1931)
Marconi, Degna. *My Father Marconi* (1957)
Marconi G. *Encyclopaedia Britannica* (13th ed) (1926)
Nalder, Major-General R. F. S. *The Royal Corps of Signals* (1950)
Nicolson, Sir Harold. *Diaries and Letters 1938–62* (3 vols) (1966–8)
Reith, J. C. W. *Broadcast over Britain* (1924)
The Marconi Company's archives
The British Broadcasting Company's written archives
Transcript of evidence given before the United States Congressional Inquiry into the *Titanic* disaster

Acknowledgements

The author wishes to thank the following for their help: the Marconi Co Ltd (Mrs B. Hance, historian), who allowed the use of many of the photographs; the British Broadcasting Corporation (Mrs M. S. Hodgson, written archives officer); the Director of the Imperial War Museum for permission to reproduce photographs; the Chief Librarian, Ministry of Defence; the County Librarian and staff, Somerset County Library; the Borough Librarian and staff, Bridgwater Borough Library; Dr Alfred N. Coldsmith of New York; and all others who gave valuable help, particularly Miss Judith Corder who so ably typed the MS.

Index

Adrianople, siege of, 91
Admiralty, British, 94, 106, 109
Aerial, Marconi directional, 77, 78
Aircraft, wireless in, 95-9
Air defence, 127-8
Air-ground WT, 95, 123-4; installations, 219-20; navigational aids, 127-8; in tanks, 132, 137; two-way, 97
Airliners, early, 135
Air Ministry, 120
Airships: British, 96, 98, 99; German, 98, 99, 126; *see also* Zeppelins
Alarm signal, 140
Alcock, John (Alcock and Brown), 134
Alexanderson alternator, 64, 65, 82, 128, 213
Alternators, radio-frequency, 82, 142
Alun Bay, station at, 22, 23
America's Cup races, 25
American Civil War, 89
American Telephone & Telegraph Co (AT & T), 69, 72, 151, 157, 186
Anglo-American Cable Co, 76, 79, 144
Aquitania, RMS, 138
Arc system, *see* Poulsen arc system
Armstrong, Edward H., 70, 71, 149
Armstrong, Neil, 206
Artillery, spotting for, 120-3
Artistes, payment of, 154
AT & T, *see* American Telephone & Telegraph Co
Audion, 68-72
Australia, RT in outback, 145-6
Auto-alarm, *see* calling device

Baird, John L., 183-5
Baird TV system, 183-5

Balkan War (1912), 90-1
Balloons, WT in, 95
Ballybunion (Ireland), station at, 162
Baltic, SS, 50, 55
Batteries, shortage of, 190
BBC, *see* British Broadcasting Company; British Broadcasting Corporation
Beam homing system, 135-6, 188, 195-6
Beam stations, commercial, 143-4
Bell, Alexander Graham, 17, 147
Bellini-Tosi direction-finder, 47-8
Bell Systems Laboratories, 72, 199
Bell Telephone Co, 148
Big Ben, 191
Biggin Hill, station at, 127
Binns, Radio Officer John R., 50, 53
Biggin Hill, station at, 127
Binns, Radio Officer John R., 50, 53
Boer War, *see* South African War
Boulogne-sur-Mer, station at, 48
Branly, Monsieur, 19, 27
Bride, Radio Officer Harold, 53
Brighton-type receiver, 146
Britain, Battle of, 191-2
British Army, 89, 91, 116-20, 130-3; BBC programmes for, 191; cavalry, 92, 116, 118, 126, 131; Desert Corps, 126; early developments, 91-3; equipment in 1914, 93, 116-17; 1st Wireless Company, 92, 95; 4th Cavalry Division, 126; inter-war period, 130-3; Palestine, 126; South African War, 89; training, insufficient, 93, 116; Western Front, 93, 117, 118
British Broadcasting Company (BBC), 166; *see also* Broadcasting
British Broadcasting Corporation (BBC),

Index

166, 176; Britain to America, 193; foreign language broadcasts, 178; German soldiers listen to, 192; 'Music While You Work', 191; relations with press, 170–3; service to army, 191; in World War II, 189–93; *see also* Broadcasting, Broadcasts, News bulletins, Television
British Field Wireless Set, 119
Broadcasting, 67, 101, 105; advertising, 152–7, 171; beamed, 177; beginnings in USA, 150 ff; controversy over, 158, 172–3, 176; a dangerous weapon, 177; decline of, 199; as entertainment, 163, 169, 176; in Germany, 177; growth of, 153, 158, 160, 194; licences (USA), 166, 168; line, 147–8; listening prohibited, 177; in national emergency, 175; to occupied countries, 177, 192–3; official attitude to, 161, 163, 172–3; from overseas (USA), 159; payment of performers, 158, 169; permitted hours, 168, 172, 175–6; as a political force, 158–9; programme material, 152–6; programmes (first BBC), 167; propaganda, 149, 177–9; religion, 148, 153; required listening, 177; to schools, 169; stations (Great Britain), 167–8; technical progress, 174–5; wavelengths (USA), 154–6
Broadcasts: air battle, 191–2; Christmas, 179; election results, 152–3, 167–8; fireside talks, 159; first, 65; first regular, 153, 162; foreign language, 178–9; historic, 180–1; from Metropolitan Opera House, 149–50; misleading, 179, 194; news, 162, 167, 170–3; nightingale, 160; from overseas (USA), 159; royal, 160, 179; sports, 153, 176, 192–3; Voice of America, 179
Brown, Whitton (Alcock and Brown), 134
Bureau des Longitudes, 42

Cable & Wireless Ltd, 144
Cable companies, competition with, 76, 79, 144
Cables, submarine, 15, 79, 80, 144
Californian, SS, 53–9
Calling devices, 24, 59, 139–41

Cape Cod (Mass.), station at, 41, 77
Carlo Alberto, Italian warship, 35, 161
Carpathia, SS, 43, 53, 59
Car radios, 159
CBS, *see* Columbia Broadcasting System
Censorship, 169, 173
Chamberlain, Neville, 189
Chelmsford, Marconi works and station at, 138, 162, 163
Chenevix-Trench, Lieut-Col, 92, 117, 126
Chester, USS, 57
Churchill, Sir Winston, 81, 98, 106, 108, 190, 198
Clapper break, 123
Clifden (Ireland), station at, 77, 80, 87
Coast stations, 37, 41
Code, 90, 106–8, 132
Coherer, 19, 32, 215–16
Columbia Broadcasting System (CBS), 157–8, 187, 193
Conductive system, 16, 17
Congestion on the air, 131; in USA, 72, 73, 154; on Western Front, 123
Conrad, Frank, 152–3
Continuous wave, 64–6, 71, 124, 130, 134
Convoy sets, 109
Coolidge, President Calvin, 158
Cotton, Radio Officer Harold T., 53, 56–8
CQD distress call, 49, 50, 53, 54
Crawford Committee, 175–6
Crimean War, 88
Crippen, Dr, 45
Crookes, Sir William, 28
Cunard Daily Bulletin, 41
CW, *see* Continuous wave
Cypher, *see* Code

Daily Express, Dublin, 23
Daily Mail, London, 45, 133, 175
Daily Telegraph, London, 44
D-day landings, 193
Defence of the Realm Act, 100, 101
De Forest, Dr Lee, 28, 31, 63, 66–70, 71, 83, 147–8
Detectors: coherer, 19, 32, 64; crystal, 35, 48, 72, 119, 125, 137, 199; electrolytic, 35, 64, 65; magnetic, 35; physiological, 218–19; valve, 69–70

Index

Detroit News, 152
DF, *see* Direction-finding
Direction-finding, 47, 110–12, 127
Direction transmission, 22, 136
Distress calls, interference with, 73
Dowding, Captain (later Air Chief Marshall), 125
Dunwoody, General, 35, 125

East Goodwin Lightship, 24, 47, 50
Eccles, Dr N. H., 119
Edison, Thomas A., 29, 63
Edison effect, 63, 66
Edward VII, King, 23, 77
Edward VIII, King, 180
'Ego' station, 154
Eiffel Tower (Paris), station at, 42, 43, 68, 115, 118, 149
Electrophone Co, 148
Elizabeth II, Queen, coronation of, 201
Empress of Ireland, RMS, 46, 60, 61
Eskimo, RMS, 48
Experimenters, amateur, 69, 72, 73, 101, 103, 142, 149, 163, 166
Exploration, use of WT in, 47

Falkland Islands, Battle of, 109, 196
Faraday, Michael, 13
'Father of Radio', *see* De Forest, Dr Lee
Fessenden, R. A., 35, 44, 64–5, 71, 147–8, 213, 217–18
Fire service, use of radio in, 202
Fisher, Admiral Sir John (later Lord), 94, 116
Flatholme, 22
Fleming, Dr J. A, 63, 83, 183
Flying Huntress, tug, 23
Forward positions, WT in, 17, 118–20, 124
Franklin, C. S., 70, 161
Frequency modulation, 71
Free speech, 159

Gardner broadcast, 191–2
General Electric Co, 151
General Strike, 173, 175–7, 179
George V, King, 159–60, 179, 180
George VI, King, 181, 185
German Army, use of wireless, 118, 124
German Navy, 109
German News Agency, 193

German overseas stations, 83, 105–6
German surrender, broadcast of, 197
Germany, broadcasting in, 177
Glace Bay (Nova Scotia), station at, 77, 80, 97, 134
Goldschmidt alternator, 65, 82, 213–14
Goonhilly Down (Cornwall) earth station, 205
Grierson, General, 96
Ground control by radio, 135, 137
Guernsey, dependence on BBC news, 192

Hankey, Sir Maurice, 79
Harding, President, 158
Hawaiian Group, stations in, 74
Hawker, Harry G., 133
Heaviside, Professor A. W., 17
Hertz, Heinrich A., 19, 27, 40, 142, 208–9
Hertzian waves, 28, 66, 75
Hertz oscillator, 20
Heterodyne system, 64, 65
Hitler, Adolph, 181
Homing beam, 135–6, 188, 195–6
Hoover, H., Secretary of Commerce (later President), 153–6, 158
Hughes, Professor D. G., 19
Hughes, Congressman J. C., 56, 58
Human voice, first on air, 65

Imperial wireless chain, 80–3, 143
Induction coil, 207–8
Inductive system, 17, 21
Interception, 90, 93, 118–19, 186
Interference, 72, 73, 122–3, 127, 164
International Communications Consortium, 205
International Conference on WT, first, 40, 149
International Ice Patrol, 46
Ionoscope, 184
Isaacs, Godfrey, 81, 165

Jackson Committee, 133
Jameson-Davis, Henry, 21
Japanese Army, use of wireless, 90
Jutland, Battle of, 110–12

Kaiser Wilhelm der Grosse, German liner, 36
KDKA station, 65, 152–3, 163

Index

Kelvin, Lord, 22
Kinescope, 184
Knickebein stations, 195
Kroonland, ss, 50
Krosigk, Count Schwerin von, 197

Lackawana Railroad, 46, 68
Ladysmith, siege of, 90
Lake Champlain, ss, 36
Land stations, 37, 84
La Panne, station at, 36
Leafield, station at, 143
Leroy, Captain P. T., 96
Lewis, Lieutenant (later Lieut-Col) Swain, 121-2
Licences: receiving, 73, 156; transmitting, 73, 154-6, 158
Liddell-Hart, Captain Basil, 132-3
Lifeboats, WT in ships', 138-9
Lindbergh, Charles, 136
Lindsay, J. B., 16
Lodge, Sir Oliver, 17, 19, 27, 28, 29, 35, 149, 211, 212
Lodge-Muirhead Syndicate, 29, 31, 211
Longitude, determination of, 43-4
Loos, Battle of, 119
Loraine, Robert, 97
Lucania, RMS, 37
Luftwaffe, 99

Madora, barge, 36
Magdeburg, German cruiser, 107-8
Magnetic detector, 35, 137, 216
Map-making, use of WT in, 44
Marconi, Annie, 20
Marconi, Degna, 21
Marconi, Guglielmo, 19-29, 35, 51, 71-8, 89, 209-11; appraisal of, 29; arrival in England, 21; directional aerial, 77, 78; learns morse, 20; Salisbury Plain experiments, 22, 142; short-wave experiments, 142; transatlantic experiments, 75-9; visit to New York, 25; Wireless Telegraph Co, 31, 81, 144; Wireless Telegraph Co of America, 31, 56, 150-1
Marconi Company, 24, 29, 36, 37, 39, 41, 48, 69, 79-82, 89, 92-4, 99, 116, 131, 137, 139, 161, 162-6
Marconi-EMI TV, 184-5
Marconi House, 129, 161, 164

Marconi 1½kW set, 35
Marconi scandal, 80-2
Marconi system, 32, 63, 64, 80, 89
Marconi Wireless and Signal Co, 22, 31
Marne, Battle of the, 123
Materials, shortages of, 113
Mauretania, RMS, 48
Maxwell, James Clerk, 18, 26
Mayday distress call, 50
Medical service, ships', 141
Meisner, A., 70, 161
Merchant marine, 100, 108, 113, 196
Metropolitan-Vickers, 164
Miniaturisation, 199-201
Misuse of wireless, 110-11, 115
Montrose, ss, 45, 46
Moon, RT from, 206
Morse, Samuel F. B., 13, 15
Murrow, Edward, 193
Music Box receiver, 152
'Music While You Work', 191

Nalder, Major-General, 120
National Broadcasting Company (NBC), 157, 158, 186
National Electric Signalling Co, 65
Nauen (Germany), station at, 83, 128
Navigation, radio aids to, 134-7, 188, 195
Navy, British, *see* Royal Navy
Navy, United States, 25, 26, 66, 68
NBC, *see* National Broadcasting Company
News, instant, 170-1, 190, 193
News bulletins, 25, 41; BBC, 162, 167, 170-3, 175-6; USA, 149
Newspapers, ships', 25, 26, 41-2
New York Herald, 25, 90
Nicolson, Sir Harold, 186, 189
Norddeich, German station at, 42
Norge, airship, 135

Olympic, RMS, 54, 56, 86

Patent Office, 32
Patents, 31, 35, 41, 68-70, 94, 137, 165-6, 206
Peace talks, use of WT in, 128-9
Pedal radio, 145-6
Pegasus, balloon, 95
Period system, 117

Philadelphia, ss, 76
Phillips, Radio Officer J. G., 53-8
Physiological detector, 218-19
Poldhu (Cornwall), station at, 41, 75, 76, 77, 134, 138
Police radio, 146, 201-3
Poole (Dorset), station at, 22, 23
Popoff, A. S., 19, 20, 27
Portable sets, 118
Postmaster-General, 165, 173
Post Office, British, 21, 22, 41, 42, 89, 102, 143, 144, 163, 171, 174
Post Office stations, 62, 84
Poulsen arc system, 68, 82, 142, 214-15
Preece, Sir William H., 17, 21
Press, attitude of, to radio, 170-3
Princesse Clementine, ss, 36
Propaganda, 105, 149, 177-9

R34, airship, 134
Radio, compulsory fitting of, 113, 135, 140
Radio, growth of, in USA, 153-4, 155, 160
Radio Act, 72, 73, 154
Radio beacons, 135
Radio Communication Co, 164
Radio Corporation of America (RCA), 31, 151, 184, 186
Radio frequency alternators, 64, 65
Radio range, 135-6
Radio relay, 148, 177
Radio Society of Great Britain, 163
Radio-sonde, 220-1
Radio telephony, 64, 65, 68, 72, 87, 100, 134, 146, 147, 161; in Australia, 145; in passenger ships, 138; on Western Front, 124
RCA, *see* Radio Corporation of America
Receivers, *see* Detectors
Reflectors, use of, 19, 22, 142
Regatta, Dublin, 23
Reith, J. C. W. (later Lord), 167, 169, 172-4, 180, 191
Religious broadcasting, 148, 153
Republic, ss, 50, 61
Resistance Units, 195
RFC, *see* Royal Flying Corps
Righi, Professor Augusto, 20
RNAS, *see* Royal Naval Air Service
Roosevelt, President F. D., 159, 186

Roosevelt, President Theodore, 77
Royal Air Force 80 Wing, 195
Royal Corps of Signals, 127
Royal Engineers, 89, 92, 95, 126
Royal Flying Corps, 95, 112, 120-6; 1st Wireless Training Unit, 123; No 9 Sqn, 122-3, 125
Royal George, ss, 48
Royal Naval Air Service, 95, 112, 125
Royal Navy, 26, 93, 94, 100, 106, 109-13, 196; Dardanelles, 112-13; shore stations, 106, 113 n, 196; WT in aircraft, 98
Royalties, 166
Rugby Radio, 145, 220
Rumanian Army, 90, 91
Running commentaries, 23, 153, 171-3, 176
Russian Army, 90, 118
Rutherford, Professor E. (later Lord), 28, 29, 35

Safety of life at sea, 24, 37, 40
Salisbury Plain, experiments on, 22, 89, 93, 94, 142
Salisbury Plain, manoeuvres on, 22, 96
Sarnoff, David, 56, 57, 150
Satellites, communication, 203-5
Scanning, invention of, 182
Scheer, Admiral, 108, 110
Secret wireless stations, 195
Selenium, 182
SHF, *see* Super High Frequency
Ship stations, 87
Short waves, use of, 138, 142
Slavonia, ss, 53
Somme, Battle of the, 123-4
SOS distress call, 50, 53, 61, 62, 140
South African War, 89, 90, 116
South Foreland Lighthouse, station at, 24, 47
Space telegraphy, 21, 28
Spark gaps, 211
Spotting for artillery, 120-3
Spy-mania, 101, 102
Stacey, Radio Officer F. S., 36-7
Sterling set, 113, 123
St John's (Newfoundland), 75, 76, 133
St Paul, ss, 25, 41
Submarines: British, 112; German, 112, 114-15

Superheterodyne circuit, 71, 174
Super High Frequency (SHF), 201
Supreme Court (USA), 70

Tanks, wireless in, 125-6, 131-3
Telefunken Co, 31, 138
Telegraph, line, 14, 15
Telemetry, *see* Radio control
Telephone, line, *see* Bell, Alexander Graham
Telephone discipline, 119
Television, 67, 182-7; Baird, 182-5; BBC service opens, 185; closure, 186; Marconi-EMI, 184-5; service resumed, 200; social impact of, 200; in USA, 182, 184, 186-7, 200
Television stations: Aelxandra Palace, 185; Crystal Palace, 183; 2TV (Harrow), 183
Telstar, 204-5
The Times, London, 55, 56, 77
Time signals, 42, 43, 44, 98, 145
Titanic disaster, 24, 37, 43, 53-9, 61
Trades Union Congress, 176
Training, inadequate, 116, 131
Trains, WT on, 46
Tramp steamers, WT on, 114-15
Transatlantic experiments, 75-9
Transatlantic flights, early, 133-4
Transatlantic radiogram, first, 77
Transatlantic radio picture service, 184
Transatlantic Times, 25
Transatlantic WT service, 77, 80, 101
Transistors, 199
Transmitters: arc, 65, 113 n, 213; CW, 64, 65, 113, 124, 130, 134, 138; Hertz, 208-9; Lodge, 211-12; Marconi, 209-11; quenched gap, 138, 211, 213; rotary gap, 138, 211, 213; short wave, 138; valve, 113 n, 143
Trawlers, WT on, 46
Trevessa, SS, 138-9
Trinity House, 24
Trowbridge, Professor E., 16, 17
Tuning, invention of, 17
Turkish Army, 90, 91

U-Boats, *see* Submarines, German
UHF, *see* Ultra High Frequency
Ultra High Frequency, 201-3
United States Navy, 25, 26, 66, 68

United Wireless Co of America, 68

Valve receivers, 119
Valves, 87, 137, 199; De Forest triode, 67, 83, 144, 150, 161; Fleming diode, 66, 83, 137, 199
Valve transmitters, 143
Very High Frequency (VHF), 146, 201-3
Vested interests, 168-73
VHF, *see* Very High Frequency
Victoria, Queen, 23
Victoria, SS, 138, 162

Walkie-talkie equipment, 146, 200, 202
Waratah, SS, 60
War Office, 89, 93, 125, 126
Washington DC, 14, 16
Wavelengths: BBC, 189; in USA, 154-6
Weather Bureau (USA), 44, 45
Weather reports, 44, 66, 164
Western Electric Co, 66, 150, 157, 164, 166
Westinghouse, 151-3, 184
White, Albert M., 66, 68
Wincanton, station at, 144
Wireless, official attitude to, 100-5, 130, 189-90
Wireless, misuse of, 110-11, 115
Wireless apparatus, illegal possession of, 100-2
Wireless silence, 108-9, 196
Wireless Society of London, 163
Wireless stations, secret, 195
Wireless Telegraph Co, Marconi's, 31, 81, 144
Wireless Telegraph Co of America, Marconi's, 31, 56, 150-1
Wireless telegraphy, invention of, 27, 28
Wireless Telegraphy Act (1904), 40
Wireless World, 47, 134, 183
World War I: end of, 129; outbreak of, 61, 82, 91, 93, 94, 142, 170
World War II 101; Big Ben, 191; British Army, 197; broadcasting, 189-93; desert army, 144; end of, 197-8; at home, 101; outbreak of, 133; at sea, 196
Wincanton, station at, 144
Writtle, station at, 163, 164

Yearbook of Wireless Telegraphy and Telephony, The (1913), 83–7, 99

Zeppelins, 99, 127–8
Zworykin, Vladimir K., 184